# Computational Physics
## An Introduction

Second Edition

# Computational Physics
## An Introduction

## Second Edition

Franz J. Vesely

*Institute of Experimental Physics*
*University of Vienna*
*Vienna, Austria*

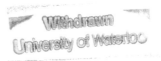
Kluwer Academic / Plenum Publishers
New York, Boston, Dordrecht, London, Moscow

Library of Congress Cataloging-in-Publication Data

Vesely, Franz.
    Computational physics: an introduction/Franz J. Vesely.—2nd ed.
      p.   cm.
    Includes bibliographical references and index.
    ISBN 0-306-46631-7
      1. Physics—Methodology.   2. Differential equations—Numerical solutions.   3. Numerical
analysis.   4. Mathematical physics.   I. Title.

QC6 .V47 2001
530.15′94—dc21

                                    2001041335

The First Edition of *Computational Physics: An Introduction* published by Plenum Press, New York, in 1994, was a translation (by the author) of *Computational Physics: Einführung in die Computative Physik*, originally published in 1993 by WUV-Universitätsverlag, Vienna, Austria.

Illustration of John von Neumann (p. 47) used with permission of Marina von Neumann Whitman (daughter of John von Neumann), Ann Arbor, Michigan.

Illustration of Ludwig Boltzmann (p. 161) used with permission of Zentralbibliothek Physik, University of Vienna, Vienna, Austria.

Illustration of Erwin Schrödinger (p. 195) used with permission of Ruth Braunizer (daughter of Erwin Schrödinger), Alpbach, Tyrol, Austria.

ISBN 0-306-46631-7

©2001 Kluwer Academic / Plenum Publishers, New York
233 Spring Street, New York, New York 10013

http://www.wkap.nl/

10  9  8  7  6  5  4  3  2  1

A C.I.P. record for this book is available from the Library of Congress

To my wife

# Preface to the Second Edition

In a rapidly evolving field such as computational physics, six years is an eternity. Even though many of the elementary techniques described here are of venerable age, their assembly into sophisticated combined methods and their intensive application to ever new problems is an ongoing and exciting process. After six years, a new edition of this textbook must therefore take into account some of the new vistas that have opened up recently.

Apart from these additions and some didactic improvements, the general structure of the book holds good. The first three chapters are devoted to a thorough, if concise, treatment of the main ingredients from numerical mathematics: *finite differences*, *linear algebra*, and *stochastics*. This exercise will prove valuable when we proceed, in chapters 4 and 5, to combine these elementary tools into powerful instruments for the integration of differential equations. The final chapters are devoted to a number of applications in selected fields: statistical physics, quantum mechanics, and hydrodynamics.

I will gradually augment this text by web-resident sample programs. These will be written in JAVA and will be accompanied by short explanations and references to this text. Thus it may prove worthwhile to pay an occasional visit to my web-site

**www.ap.univie.ac.at/users/Franz.Vescly/**

to see if any new applets have sprung up.

Vienna, August 2000

**A subjective view on related texts:**

In the stalls of university bookstores numerous texts may be found that have "Computational Physics" in their title. Being a competitor I will not attempt at an objective critique of them, but before you choose one I suggest you answer two questions for yourself: 1) Is it really a book on computational physics, or rather a treatise on the author's own branch of physics, with special consideration of numerical techniques? 2) Are the algorithms used in the book thoroughly explained and derived in a compelling way, or are they simply introduced as "falling out of the blue"? – If the book passes both tests, take it.

After this general caveat I feel free to recommend the following volumes, as either similar in spirit or complementary to mine. Anything that is not mentioned here need not be bad – I simply may not be aware of it.

POTTER, D.: COMPUTATIONAL PHYSICS. Wiley, New York 1980.
*Very valuable text; in some places too demanding for the beginner.*

HOCKNEY, R. W., AND EASTWOOD, J. W.: COMPUTER SIMULATION USING PARTICLES. McGraw-Hill, New York 1981.
*Very good, particularly, but not exclusively, for plasma physicists; covers large areas of computational physics, in spite of the seemingly restrictive title.*

HOOVER, W. G.: COMPUTATIONAL STATISTICAL MECHANICS. Elsevier, Amsterdam 1991.
*Beautiful account of how to do profound physics by computing.*

PRESS, W. H., FLANNERY, B. P., TEUKOLSKY, S. A., AND VETTERLING, W. T.: NUMERICAL RECIPES IN FORTRAN. Cambridge University Press, Cambridge 1992.
*Excellent handbook of modern numerical mathematics; comes with sample programs in various programming languages.*

GIORDANO, N. J.: COMPUTATIONAL PHYSICS. Prentice-Hall, New Jersey 1997.
*This is one of those texts in which little is said about the origin of the the algorithms used. However, it is redeemed by its large collection of charming physical applications. Use it together with a more method-oriented text.*

GOULD, H., AND TOBOCHNIK, J.: INTRODUCTION TO COMPUTER SIMULATION METHODS: APPLICATION TO PHYSICAL SYSTEMS. Addison-Wesley, Reading 1996.
*Nice "hands-on" introduction; starts out with elementary physics problems and works up to such cutting-edge applications as dynamical quantum simulation and renormalization.*

GARCIA, A. L.: NUMERICAL METHODS FOR PHYSICS. Prentice Hall, New Jersey, 1999.
*Carefully organized introduction to the field; presents many examples, including code and graphics.*

GERSHENFELD, N.: THE NATURE OF MATHEMATICAL MODELING. Cambridge University Press, Cambridge 1999.
*Grand tour through applied mathematics, covering analytical, numerical and observational models.*

# Preface to the First Edition

Computational physics is physics done by means of computational methods. Computers do not enter into this tentative definition. A number of fundamental techniques of our craft were introduced by Newton, Gauss, Jacobi, and other pioneers who lived quite some time before the invention of workable calculating machines. To be sure, nobody in his or her right state of mind would apply stochastic methods by throwing dice, and the iterative solution of differential equations is feasible only in conjunction with the high computing speed of electronic calculators. Nevertheless, computational physics is much more than "Physics Using Computers."

*The essential point in computational physics is not the use of machines, but the systematic application of numerical techniques in place of, and in addition to, analytical methods, in order to render accessible to computation as large a part of physical reality as possible.*

In all quantifying sciences the advent of computers rapidly extended the applicability of such numerical methods. In the case of physics, however, it triggered the evolution of an entirely new field with its own goals, its own problems, and its own heroes. Since the late forties, computational physicists have developed new numerical techniques (Monte Carlo and molecular dynamics simulation, fast Fourier transformation), discovered unexpected physical phenomena (Alder vortices, shear thinning), and posed new questions to theory and experiment (chaos, strange attractors, cellular automata, neural nets, spin glasses, ... ).

An introductory text on computational physics must first of all provide the basic numerical/computational techniques. This will be done in Parts I and II. These chapters differ from the respective treatments in textbooks on numerical mathematics in that they are less comprehensive – only those methods that are of importance in physics will be described – and in focusing more on "recipes" than on stringent proofs.

Having laid out the tools, we may then go on to explain specific problems of computational physics. Part III provides a – quite subjective – selection of modern fields of research. A systematic classification of applied computational physics is not possible, and probably not even necessary. In fact, *all* areas of physics have been fertilized, and to some extent transformed, by the massive (and intelligent) use of numerical methods. Any more advanced sequels to this introductory book would therefore have to be either collections of contributions by several authors, or else monographs on various subfields of computational physics.

Appendix A is devoted to a short description of some properties of computing machines. In addition to those inaccuracies and instabilities that are inherent in the numerical methods themselves, we always have to keep in mind the sources of error that stem from the finite accuracy of the internal representation of numbers in a computer.

In Appendix B an outline of the technique of "Fast Fourier Transformation" (FFT) is given. The basic properties and the general usefulness of the Fourier transform need no explanation, and its discretized version is easy to understand. But what about the practical implementation? By simply "coding along" we would end up at an impasse. The expense in computing time would increase as the square of the number $N$ of tabulated values of the function to be transformed, and things would get sticky above $N = 500$ or so. A trick that is usually ascribed to the authors Cooley and Tukey (see [PRESS 86]) leads to a substantial acceleration that only renders the procedure practicable. In this fast method, the computing time increases as $N \log_2 N$ only, so that table lengths of the order $N = 10.000$ are no problem at all.

When pregnant with a book, one should avoid people. If, however, one has to seek them, be it to ask for advice, to request support or to beg for the taking over of teaching loads, they should be such patient and helpful people like Renato Lukač, Martin Neumann, Harald Posch, Georg Reischl or Konrad Singer.

What one is doing to one's family cannot be made good by words alone.

Vienna, March 1993                                         F. J. Vesely

# Contents

# Computational Physics
An Introduction

Second Edition

# Part I

# The Three Pillars of Computational Physics

Most of the methods used by computational physicists are drawn from three areas of numerical mathematics, namely from the *calculus of differences*, from *linear algebra*, and from *stochastics*.

*Difference calculus:* Here we use finite differences, as opposed to infinitesimal differentials, as the elements of computation. Let $f(x)$ be some function of a single variable. In standard calculus, at least the independent variable $x$ is assumed to vary in a continuous manner. Whenever $x$ is limited to a discrete set of values $x_k$ ($k = 1, 2, ...$), we are entering the realm of finite differences.

History took the opposite route. "Divided differences" of the form $(f_{k+1} - f_k)/(x_{k+1} - x_k)$ served as the base camp when Newton and Leibniz set out to attack the summit of infinitesimal calculus. But as soon as the frontier towards infinitely small quantities had been crossed, and the rules of the differential and integral calculus had been established, physicists grew ever more enthralled by these miraculous new tools. The calculus of infinitesimals became a "hit", much like the computer did in our days. And much like the computer, it acted to focus the attention of physicists on those problems that could most readily be tackled with this apparatus. Other topics were shelved for a while, and in the course of many generations were almost forgotten.

A striking example for this selectivity of scientific perception may be found in Kepler's problem. By applying the methods of calculus to the equations of motion of two gravitating celestial bodies we may eventually come up with analytical expressions for the trajectories. For three or more interacting bodies this is in general impossible. And so it comes that every student of physics very soon learns how to solve the two-body problem by analytical means, whereas the study of three and more-body problems became the task of an exclusive circle of specialists. Only in recent years the re-encounter with chaos and incomputability in deterministic mechanics helped physicists to become once more aware of the wealth of phenomena dwelling beyond the "zoo of pure cases."

The methods of difference calculus, which are older and more clumsy than those of differential calculus, yet remain applicable even in the case of three, four, or hundreds of interacting bodies. And we are not even restricted to the $1/r$ - interaction of gravitating masses. The price we have to pay for this greater freedom in the selection of mechanical problems is the fact that we can no more obtain a closed formula for the trajectories, but only a – albeit arbitrarily fine – table of trajectory points.

It is quite understandable that in the three centuries since the publication of the "Principia" this more practical aspect of Newton's work was somewhat neglected. The repetitive application of iterative algorithms is time-consuming and tedious; a renaissance of this branch of computational physics could take place only after the development of efficient computing machinery. In its modern version it is known as classical-mechanical simulation, or – in a specific context – as "molecular

4

dynamics" simulation.

*Linear algebra* is the second tributary to our methodological pool. Of course, any attempt of a comprehensive coverage of this field would go far beyond the frame of this text. However, the matrices that are of importance in computational physics very often have a rather simple structure. For example, by employing the finite difference formalism to convert a partial differential equation into a system of linear equations, we end up with a matrix of coefficients that has its non-zero elements concentrated near the main diagonal – i.e. a "diagonally dominated" matrix. And in the framework of stochastic methods we encounter covariance matrices which are always symmetric, real, and positive definite.

We will therefore concentrate on those techniques that have special importance in computational physics. Just for completeness, a short survey of the standard methods for the exact solution of linear systems of equations will be given. The main part of Chapter 2, however, will be devoted to procedures that are particularly suited for symmetric real matrices and to those iterative methods that converge particularly fast when applied to diagonally dominated matrices. There are also iterative techniques for determining eigenvalues and eigenvectors which may be applied in addition to or in place of exact methods.

*Stochastics* is statistics turned upside down. Textbooks on statistics are in general concerned with procedures that allow us to find and quantify certain regularities in a given heap of numbers. Contrariwise, in stochastics these statistical properties are given beforehand, and an important task then is the production of "random numbers" with just those properties.

In contrast to the other two pillars of computational physics, stochastics is a product of the computer age. In the forties, after the still rather failure-prone ENIAC, the MANIAC was constructed as the second fully electronic computing machine. (Incidentally, Nicholas Metropolis, who hated this kind of abbreviations, had meant to bring the custom to an end once and for all by introducing a particularly idiotic acronym [COOPER 89]; the further history of Computerspeak, from UNIVAC to WYSIWYG, is proof of the grandiose failure of his brave attempt.)

The primary use of these early computers was numerical neutron physics. The transport of neutrons through an inhomogeneous medium, be it an atomic bomb or the core of a reactor, is described by complicated integro-differential equations. Instead of solving these transport equations directly, Metropolis, Fermi, Ulam and others [ULAM 47, METROPOLIS 49] used a stochastic procedure which they dubbed "Monte Carlo method." They programmed their machine in such a way that it sampled at random many individual neutron paths. A neutron would be sent on its way with a typical velocity, could penetrate more or less deeply into the material, was then absorbed or scattered into some new direction, and so on. By taking the average over many such neutron trajectories one could determine the mean flux at some given point.

A similar idea is the basis of the method of "Brownian dynamics." Here the random motion of mesoscopic particles is simulated according to a simple rule. Small, randomly sampled path increments are combined to a trajectory that closely resembles the typical random walk of Brownian diffusors. By adding external conditions, such as absorbing walls or force fields, one may simulate non-trivial, physically relevant situations.

For the evaluation of thermodynamic averages we may use the statistical-mechanical Monte Carlo method, which at first sight bears little resemblance to its namesake in neutron physics. Here, the high-dimensional canonical phase space of an $N$-particle system is perambulated by random steps. By a sophisticated trick that is again due to Metropolis, we can achieve that phase space regions with a large Boltzmann factor will be visited more frequently than regions with small thermodynamic probability. Thus it is possible to determine canonical averages by simply taking mean values over such random walks.

A surprise bounty was discovered in the Eighties. It turned out that the basic principle of the Monte Carlo method can be of great value even outside of statistical mechanics. If the temperature is slowly decreased during the random walk through phase space, eventually only the regions with lowest energy will be visited. With a bit of luck we will end up not in some local energy dip, but in the global minimum. This means that we have here a stochastic method for locating the minimum of a quantity (the energy) depending on a large number of variables (the $3N$ particle coordinates.) Such notoriously difficult multidimensional minimization problems are to be found in many branches of science. Applications of this "Simulated annealing" technique range from the optimization of printed circuits on computer chips to the analysis of complex neural nets.

Yet another group of stochastic optimization techniques is known as "Genetic Algorithms". As the name implies, they roughly mimick biological adaptation to locate the minimum of a multivariable function.

The three main methodological sources of computational physics will be treated in detail in the three Chapters of Part I. It is not my ambition to prove each and every formula in full mathematical rigor. More often than not we will content ourselves with arguments of plausibility or with citations, if only we end up with a concrete algorithm or procedure.

# Chapter 1

# Finite Differences

*Isaac Newton handled differences and differentials with equal prowess*

Let $f(x)$ be a continuous function of one variable. The values of this function are given only for discrete, and equidistant, values of $x$:

$$f_k \equiv f(x_k), \quad \text{where} \quad x_k \equiv x_0 + k\,\Delta x \tag{1.1}$$

The quantity

$$\Delta f_k \equiv f_{k+1} - f_k \tag{1.2}$$

is called "forward difference" at the point $x_k$. By repeated application of this definition we obtain the higher forward differences

$$\Delta^2 f_k \equiv \Delta f_{k+1} - \Delta f_k = f_{k+2} - 2f_{k+1} + f_k\,, \tag{1.3}$$

$$\Delta^3 f_k \equiv \Delta^2 f_{k+1} - \Delta^2 f_k = f_{k+3} - 3f_{k+2} + 3f_{k+1} - f_k \tag{1.4}$$

$$\vdots$$

The coefficients of the terms $f_l$ are just the binomial coefficients which may conveniently be taken off Pascal's triangle. Quite generally, we have

$$\Delta^r f_k \equiv \sum_{i=0}^{r} (-1)^i \binom{r}{i} f_{k+r-i} \tag{1.5}$$

For given $\Delta x$, the values of $\Delta f_k$ provide a more or less accurate measure of the slope of $f(x)$ in the region towards the right of $x_k$. Similarly, the higher forward differences are related to the higher derivatives of $f(x)$ in that region.

The "backward difference" at $x_k$ is defined as

$$\nabla f_k \equiv f_k - f_{k-1} \tag{1.6}$$

and the higher backward differences are

$$\nabla^2 f_k \equiv \nabla f_k - \nabla f_{k-1} = f_k - 2f_{k-1} + f_{k-2} \tag{1.7}$$

etc., or, in general

$$\nabla^r f_k \equiv \sum_{i=0}^{r} (-1)^i \binom{r}{i} f_{k-i} \tag{1.8}$$

In the formulae given so far only table values to the right or to the left of $x_k$ were used. In contrast, the definition of the "central difference" is symmetric with respect to $x_k$:

$$\delta f_k \equiv f_{k+1/2} - f_{k-1/2} \tag{1.9}$$

At first sight this definition does not look all too useful, since by our assumption only table values of $f_k$ with integer indices $k$ are given. However, if we go on to higher central differences, we find that at least the differences of *even order* contain only terms with integer indices:

$$\begin{aligned}
\delta^2 f_k &= f_{k+1} - 2f_k + f_{k-1} & (1.10)\\
\delta^3 f_k &= f_{k+3/2} - 3f_{k+1/2} + 3f_{k-1/2} - f_{k-3/2} & (1.11)\\
\delta^4 f_k &= f_{k+2} - 4f_{k+1} + 6f_k - 4f_{k-1} + f_{k-2} & (1.12)
\end{aligned}$$

and in general

$$\delta^r f_k \equiv \sum_{i=0}^{r} (-1)^i \binom{r}{i} f_{k+r/2-i} \tag{1.13}$$

One final definition, which will serve primarily to provide access to the odd-order central differences, pertains to the "central mean",

$$\begin{aligned}
\mu f_k &\equiv \frac{1}{2}[f_{k+1/2} + f_{k-1/2}] & (1.14)\\
\mu^2 f_k &\equiv \frac{1}{2}[\mu f_{k+1/2} + \mu f_{k-1/2}]\\
&= \frac{1}{4}[f_{k+1} + 2f_k + f_{k-1}] & (1.15)
\end{aligned}$$

etc.

In place of a – not obtainable – central difference of odd order, such as $\delta f_k$, we may then use the central mean of this difference, namely

$$\mu\,\delta f_k \quad \equiv \quad \frac{1}{2}[\delta f_{k+1/2} + \delta f_{k-1/2}]$$

$$= \quad \frac{1}{2}[f_{k+1} - f_{k-1}] \qquad\qquad (1.16)$$

which again contains only table values that are known.

# 1.1    Interpolation Formulae

*Nota bene*: this section is not concerned with "interpolation" – in that case we would have to rehearse basic numerical skills like spline, Aitken or other interpolation techniques – but with the derivation of interpolation *formulae* which may further on be used as formal expressions. We will later differentiate them (Section 1.2), integrate them (Chapter 4) and insert them in systems of linear equations (Chapter 5).

So far we have not made use of the assumption that the points $x_k$ are arranged in regular intervals; the relations following now are valid only for equidistant table points. This assumption of constant step width may seem restrictive. However, in computational physics our aim is in general not to interpolate within some given – and certainly not always conveniently spaced – tables. Rather, the following interpolation formulae shall serve us as a basis for the derivation of iterative algorithms to solve differential equations. In other words, we will develop methods to *produce*, on the grounds of a given physical law, a sequence of "table values." This implies that as a rule we have the freedom to assume some fixed step width.

Thus, let $\Delta x$ be constant, and let $x_k$ be some particular point in the table $\{x_k, f_k;\ k = 1, 2, ..\}$. As a measure for the distance between an arbitrary point on the $x$-axis and the point $x_k$ we will use the normalized quantity

$$u \equiv \frac{x - x_k}{\Delta x} \qquad\qquad (1.17)$$

## 1.1.1    NGF Interpolation

We can obtain an interpolation approximation $F_m(x)$ to the tabulated function by threading a polynomial of order $m$ through $m + 1$ table points. If we use only points to the *right* of $x_k$ (and $x_k$ itself), the general polynomial approximation may be written in terms of *forward* differences as follows:

**NGF interpolation:**

$$F_m(x) = f_k + \binom{u}{1}\Delta f_k + \binom{u}{2}\Delta^2 f_k + \dots$$

$$= f_k + \sum_{l=1}^{m}\binom{u}{l}\Delta^l f_k + O[(\Delta x)^{m+1}] \tag{1.18}$$

where

$$\binom{u}{l} \equiv \frac{u(u-1)\dots(u-l+1)}{l!} \tag{1.19}$$

The expression 1.18 is known as the Newton-Gregory/forward or NGF interpolation formula.

The remainder term in 1.18 requires a grain of salt. Strictly speaking, this error term has the form

$$R = O\left[f^{(m+1)}(x')\frac{(x-x')^{m+1}}{(m+1)!}\right] \tag{1.20}$$

where $x = x'$ denotes the position of the maximum of $|f^{(m+1)}(x)|$ in the interval $[x_k, x_{k+m}]$. Putting

$$x - x' \equiv \xi\,\Delta x \tag{1.21}$$

we have

$$R = O\left[f^{(m+1)}(x')\frac{\xi^{m+1}}{(m+1)!}(\Delta x)^{m+1}\right] \tag{1.22}$$

By the simpler notation $O[(\Delta x)^{m+1}]$ we only want to stress which power of $\Delta x$ is relevant for the variation of the remainder term. The other factors in the remainder are assumed to be harmless. This is to say, the function to be approximated should be continuous and differentiable, and $x$ should be situated, in the case of extrapolation, not too far from the interval $[x_k, x_{k+m}]$.

EXAMPLE: Taking $m = 2$ in the general NGF formula (1.18) we obtain the parabolic approximation

$$F_2(x) = f_k + \frac{\Delta f_k}{\Delta x}(x - x_k) + \frac{1}{2}\frac{\Delta^2 f_k}{(\Delta x)^2}(x - x_k)(x - x_{k+1}) + O[(\Delta x)^3] \tag{1.23}$$

## 1.1.2   NGB Interpolation

We obtain the Newton-Gregory/backward (or NGB) formula if we use, in setting up the polynomial, only table values at $x_k, x_{k-1}, \dots$:

**NGB interpolation:**

$$F_m(x) \;=\; f_k + \frac{u}{1!}\nabla f_k + \frac{u(u+1)}{2!}\nabla^2 f_k + \dots$$

$$\;=\; f_k + \sum_{l=1}^{m} \binom{u+l-1}{l}\nabla^l f_k + O[(\Delta x)^{m+1}] \qquad (1.24)$$

EXAMPLE: With $m = 2$ we arrive at the parabolic NGB approximation

$$F_2(x) = f_k + \frac{\nabla f_k}{\Delta x}(x - x_k) + \frac{1}{2}\frac{\nabla^2 f_k}{(\Delta x)^2}(x - x_k)(x - x_{k-1}) + O[(\Delta x)^3] \qquad (1.25)$$

## 1.1.3    ST Interpolation

By "Stirling" (or ST) interpolation we denote the formula we obtain by employing the central differences $\delta f_k$, $\delta^2 f_k$ etc. Here we are faced with the difficulty that central differences of odd order cannot be evaluated using a given table of function values. Therefore we replace each term of the form $\delta^{2l+1} f_k$ by its central mean. In this manner we obtain a "symmetrical" formula in which the table points $x_k, x_{k+1}, \dots x_{k\pm n}$ are used to construct a polynomial of even order $m = 2n$:

**ST interpolation:**

$$F_{2n}(x) \;=\; f_k + u\mu\delta f_k + \frac{u^2}{2!}\delta^2 f_k + \frac{u^3 - u}{3!}\mu\delta^3 f_k + \frac{u^4 - u^2}{4!}\delta^4 f_k + \dots$$

$$\;=\; f_k + \sum_{l=1}^{n} \binom{u+l-1}{2l-1}\left[\mu\delta^{2l-1} f_k + \frac{u}{2l}\delta^{2l} f_k\right]$$

$$+ O[(\Delta x)^{2n+1}] \qquad (1.26)$$

EXAMPLE: Setting $n = 1$ (or $m = 2$) in 1.26 yields the parabolic Stirling formula

$$F_2(x) = f_k + \frac{\mu\delta f_k}{\Delta x}(x - x_k) + \frac{1}{2}\frac{\delta^2 f_k}{(\Delta x)^2}(x - x_k)^2 + O[(\Delta x)^3] \qquad (1.27)$$

Within a region symmetric about $x_k$ the Stirling polynomial gives, for equal orders of error, the "best" approximation to the tabulated function. (The "goodness" of an approximation, which will not be explained any further, has to do with the maximum value of the remainder term in the given interval.)

It is in keeping with the uncommunicative style of Isaac Newton that he permitted his *"regula quae ad innumera aequalia intervalla recte se habet, quia tum recte se habebit in locis intermediis"* [NEWTON 1674] to be published only in the year 1711 [JONES 1711], although he had found it, as is evident from various manuscripts, letters and the "Principia ...", no later than 1675-76. (Incidentally, the immediate occasion for his early involvement with the interpolation problem was the request of a private scholar by the name of John Smith, who had undertaken to publish an exact table of square, cubic and quartic roots of the numbers 1 to 10.000.) As a consequence of this reluctance, various special forms of Newton's formulae are ascribed to Gregory, Cotes, Bessel and Stirling, although these authors as a rule would respectfully point out Newton's priority.

## 1.2    Difference Quotients

Thanks to Newton, Gregory and Stirling we are now in possession of a continuous and several times differentiable function which at least at the table points coincides with the given function. Whether it does so in between these points we cannot know – it is just our implicit hope. But now we go even further in our optimism. The *derivative* of a function that is given only at discrete points is not known even at these points. Nevertheless we will assume that the derivatives of our interpolation polynomial are tolerably good approximations to those unknown differential quotients. The procedure of approximating derivatives by difference quotients has recently come to be termed "differencing."

In order to be able to differentiate the various polynomials, 1.18, 1.24 and 1.26, we have to consider first how to differentiate terms of the form 1.19 with respect to $u$. The first two derivatives of such generalized binomial coefficients are

$$\frac{d}{du}\binom{u}{l} = \binom{u}{l}\sum_{i=0}^{l-1}\frac{1}{u-i} \tag{1.28}$$

and

$$\frac{d^2}{du^2}\binom{u}{l} = \begin{cases} 0 & \text{for } l = 1 \\ \binom{u}{l}\sum_{i=0}^{l-1}\sum_{\substack{j=0 \\ j\neq i}}^{l-1}\frac{1}{(u-i)(u-j)} & \text{for } l \geq 2 \end{cases} \tag{1.29}$$

### 1.2.1    DNGF Formulae

Using the above expressions in differentiating the NGF polynomial 1.18, we find for the first two derivatives in a small region – preferably towards the right – around $x_k$:

$$F_m'(x) = \frac{1}{\Delta x}\sum_{l=1}^{m}\Delta^l f_k\binom{u}{l}\sum_{i=0}^{l-1}\frac{1}{u-i} + O[(\Delta x)^m] \tag{1.30}$$

---

**DNGF:**

$$F_m{}'(x_k) = \frac{1}{\Delta x}\left[\Delta f_k - \frac{\Delta^2 f_k}{2} + \frac{\Delta^3 f_k}{3} - \frac{\Delta^4 f_k}{4} + \ldots\right]$$

$$= \frac{1}{\Delta x}\sum_{l=1}^{m}(-1)^{l-1}\frac{\Delta^l f_k}{l} + O[(\Delta x)^m] \qquad (1.32)$$

**DDNGF:**

$$F_m{}''(x_k) = \frac{1}{(\Delta x)^2}\left[\Delta^2 f_k - \Delta^3 f_k + \frac{11}{12}\Delta^4 f_k - \ldots\right]$$

$$= \frac{2}{(\Delta x)^2}\sum_{l=2}^{m}(-1)^l\frac{\Delta^l f_k}{l}\sum_{i=1}^{l-1}\frac{1}{i} + O[(\Delta x)^{m-1}] \qquad (1.33)$$

Table 1.1: NGF approximations to the first and second derivatives at the point $x_k$

$$F_m{}''(x) = \frac{1}{(\Delta x)^2}\sum_{l=2}^{m}\Delta^l f_k\binom{u}{l}\sum_{i=0}^{l-1}\sum_{\substack{j=0\\j\neq i}}^{l-1}\frac{1}{(u-i)(u-j)} + O[(\Delta x)^{m-1}]\,(1.31)$$

We can see that the quality of the approximation, as given by the order of the remainder term, has suffered somewhat; the order of the error has decreased by 1 and 2, respectively.

In the numerical treatment of differential equations we will not need the differentiated interpolation formulae in their full glory. It will be sufficient to know $F'(x)$ and $F''(x)$ at the supporting points of the grid, in particular at the point $x = x_k$, i.e. for $u = 0$. The relevant expressions are listed in table 1.1.

EXAMPLE: Taking $m = 2$ we obtain as the DNGF approximation to the first derivative at $x = x_k$:

$$F_2{}'(x_k) = \frac{1}{\Delta x}\left[\Delta f_k - \frac{\Delta^2 f_k}{2}\right] + O[(\Delta x)^2]$$

$$= \frac{1}{\Delta x}\left[-\frac{1}{2}f_{k+2} + 2f_{k+1} - \frac{3}{2}f_k\right] + O[(\Delta x)^2] \qquad (1.34)$$

**DNGB:**

$$F_m{}'(x_k) = \frac{1}{\Delta x}\left[\nabla f_k + \frac{\nabla^2 f_k}{2} + \frac{\nabla^3 f_k}{3} + \frac{\nabla^4 f_k}{4} + \cdots\right]$$

$$= \frac{1}{\Delta x}\sum_{l=1}^{m}\frac{\nabla^l f_k}{l} + O[(\Delta x)^m] \qquad (1.37)$$

**DDNGB:**

$$F_m{}''(x_k) = \frac{1}{(\Delta x)^2}\left[\nabla^2 f_k + \nabla^3 f_k + \frac{11}{12}\nabla^4 f_k + \cdots\right]$$

$$= \frac{2}{(\Delta x)^2}\sum_{l=2}^{m}\frac{\nabla^l f_k}{l}\sum_{i=1}^{l-1}\frac{1}{i} + O[(\Delta x)^{m-1}] \qquad (1.38)$$

Table 1.2: NGB approximations to the first and second derivatives at $x_k$

## 1.2.2  DNGB Formulae

Of course, we can play the same game using the NGB interpolation polynomial. By twice differentiating equ. 1.24 we find the expressions

$$F_m{}'(x) = \frac{1}{\Delta x}\sum_{l=1}^{m}\nabla^l f_k\binom{u+l-1}{l}\sum_{i=0}^{l-1}\frac{1}{u+i} + O[(\Delta x)^m] \qquad (1.35)$$

$$F_m{}''(x) = \frac{1}{(\Delta x)^2}\sum_{l=2}^{m}\nabla^l f_k\binom{u+l-1}{l}\sum_{i=0}^{l-1}\sum_{\substack{j=0\\ j\neq i}}^{l-1}\frac{1}{(u+i)(u+j)}$$

$$+ O[(\Delta x)^{m-1}] \qquad (1.36)$$

which work best when applied to the left of $x_k$. In particular, at the position $x = x_k$, which means taking $u = 0$, we find the expressions listed in table 1.2.

EXAMPLE: $m = 2$ yields

$$F_2{}'(x_k) = \frac{1}{\Delta x}\left[\nabla f_k + \frac{\nabla^2 f_k}{2}\right] + O[(\Delta x)^2]$$

$$= \frac{1}{\Delta x}\left[\frac{3}{2}f_k - 2f_{k-1} + \frac{1}{2}f_{k-2}\right] + O[(\Delta x)^2] \qquad (1.39)$$

---

**DST:**

$$F_{2n}{}'(x_k) = \frac{1}{\Delta x}\left[\mu\delta f_k - \frac{1}{6}\mu\delta^3 f_k + \frac{1}{30}\mu\delta^5 f_k - \frac{1}{140}\mu\delta^7 f_k + \dots\right]$$

$$= \frac{1}{\Delta x}\sum_{l=1}^{n}\mu\delta^{2l-1} f_k(-1)^{l-1}\frac{[(l-1)!]^2}{(2l-1)!} + O[(\Delta x)^{2n}] \qquad (1.42)$$

**DDST:**

$$F_{2n}{}''(x_k) = \frac{1}{(\Delta x)^2}\left[\delta^2 f_k - \frac{1}{12}\delta^4 f_k + \frac{1}{90}\delta^6 f_k - \frac{1}{560}\delta^8 f_k + \dots\right]$$

$$= \frac{1}{(\Delta x)^2}\sum_{l=1}^{n}\delta^{2l} f_k\frac{(-1)^{l-1}}{l}\frac{[(l-1)!]^2}{(2l-1)!} + O[(\Delta x)^{2n}] \qquad (1.43)$$

Table 1.3: Stirling approximants to the first and second derivatives at the point $x_k$

### 1.2.3   DST Formulae

Lastly, we may choose to differentiate the Stirling formula 1.26 once and twice; it is to be expected that the expressions obtained in this manner will function best in an interval that is centered around $x_k$:

$$F_{2n}{}'(x) =$$
$$\frac{1}{\Delta x}\sum_{l=1}^{n}\binom{u+l-1}{2l-1}\left\{\left[\mu\delta^{2l-1} f_k + \frac{u}{2l}\delta^{2l} f_k\right]\sum_{i=1}^{2l-1}\frac{1}{u-l+i} + \frac{1}{2l}\delta^{2l} f_k\right\}$$
$$+ O[(\Delta x)^{2n}] \qquad (1.40)$$

$$F_{2n}{}''(x) = \frac{\delta^2 f_k}{(\Delta x)^2} + \frac{1}{(\Delta x)^2}\sum_{l=2}^{n}\binom{u+l-1}{2l-1}$$
$$\cdot\left\{\left[\mu\delta^{2l-1} f_k + \frac{u}{2l}\delta^{2l} f_k\right]\sum_{i=1}^{2l-1}\sum_{\substack{j=1\\\neq i}}^{2l-1}\frac{1}{(u-l+i)(u-l+j)} + \frac{\delta^{2l} f_k}{l}\sum_{i=1}^{2l-1}\frac{1}{u-l+i}\right\}$$
$$+ O[(\Delta x)^{2n-1}] \qquad (1.41)$$

At $x = x_k$ (i.e. $u = 0$) we find the formulas given in table 1.3.

<u>EXAMPLE:</u> $n = 1$ yields for the first derivative the approximation

$$F_2{}'(x_k) = \frac{1}{\Delta x}[\mu\delta f_k] + O[(\Delta x)^2]$$

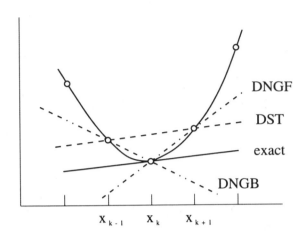

Figure 1.1: Comparison of various simple approximations to the first differential quotient

$$= \frac{1}{2\Delta x}[f_{k+1} - f_{k-1}] + O[(\Delta x)^2] \qquad (1.44)$$

The particular efficiency of the Stirling formulae is illustrated by the fact that by including just the first term on the right-hand side of 1.42 we already have an approximation of first order – in the case of NGF and NGB, inclusion of the first terms alone yields only zero-order approximations (see Figure 1.1):

$$DNGF : F'(x_k) = \frac{\Delta f_k}{\Delta x} + O[\Delta x] = \frac{1}{\Delta x}[f_{k+1} - f_k] + O[\Delta x] \qquad (1.45)$$

$$DNGB : F'(x_k) = \frac{\nabla f_k}{\Delta x} + O[\Delta x] = \frac{1}{\Delta x}[f_k - f_{k-1}] + O[\Delta x] \qquad (1.46)$$

$$DST : F'(x_k) = \frac{\mu\delta f_k}{\Delta x} + O[(\Delta x)^2] = \frac{1}{2\Delta x}[f_{k+1} - f_{k-1}]$$
$$+ O[(\Delta x)^2] \qquad (1.47)$$

Furthermore it should be noted that the remainder in 1.43 is of order $2n$. From 1.41 one would expect $2n - 1$, but by a subtle cancellation of error terms only the orders $2n$ and higher survive when we put $u = 0$. This is one reason why symmetric formulae such as those of the Stirling family are generally superior to asymmetric ones. It will turn out that the Stirling approximation to the second differential quotient serves particularly well in the numerical treatment of differential equations of second order. Keeping in mind the very special role such second-order differential equations play in physics, we regard the following formula with some respect and great expectation:

$$DDST : F''(x_k) = \frac{\delta^2 f_k}{(\Delta x)^2} + O[(\Delta x)^2]$$

$$= \frac{1}{(\Delta x)^2}[f_{k+1} - 2f_k + f_{k-1}] + O[(\Delta x)^2] \qquad (1.48)$$

The sample application given in 1.4.1 will show that our high hopes are justified.

## 1.3    Finite Differences in Two Dimensions

So far we have considered functions that depend on one variable only. However, the above definitions and relations may easily be generalized to two or more independent variables. As an example, let $f(x, y)$ be given for equidistant values of $x$ and $y$, respectively:

$$f_{i,j} \equiv f(x_0 + i\,\Delta x, y_0 + j\,\Delta y). \qquad (1.49)$$

We will use the short notation

$$f_x \equiv \frac{\partial f(x, y)}{\partial x} \qquad (1.50)$$

*et mut.  mut.* for the partial derivatives of the function $f$ with respect to its arguments.

For the numerical treatment of partial differential equations we again have to "difference", i.e. to construct discrete approximations to the partial derivatives at the base points $(x_i, y_j)$. As before, there are several possible ways to go about it, each of them related to one of the various approximations given above. Using the DNGF-, the DNGB- or the DST-approximation of lowest order, we have

$$[f_x]_{i,j} \approx \frac{1}{\Delta x}[f_{i+1,j} - f_{i,j}] + O[\Delta x] \equiv \frac{\Delta_i f_{i,j}}{\Delta x} + O[\Delta x] \qquad (1.51)$$

or

$$[f_x]_{i,j} \approx \frac{1}{\Delta x}[f_{i,j} - f_{i-1,j}] + O[\Delta x] \equiv \frac{\nabla_i f_{i,j}}{\Delta x} + O[\Delta x] \qquad (1.52)$$

or

$$[f_x]_{i,j} \approx \frac{1}{2\Delta x}[f_{i+1,j} - f_{i-1,j}] + O[(\Delta x)^2] \equiv \frac{\mu\delta_i f_{i,j}}{\Delta x} + O[(\Delta x)^2] \qquad (1.53)$$

Again, the simple insertion of the *central* difference quotient in place of the derivative results in an order of error that is higher by 1 than if we use either of the other finite difference expressions.

The next step is the approximation of the *second* derivative of $f(x, y)$ by difference quotients. By again fixing one of the independent variables – $y$, say – and considering only $f_{xx}$, we obtain, in terms of the Stirling (centered) approximation,

$$[f_{xx}]_{i,j} \approx \frac{1}{(\Delta x)^2}[f_{i+1,j} - 2f_{i,j} + f_{i-1,j}] + O[(\Delta x)^2]$$

$$\equiv \frac{\delta_i^2 f_{i,j}}{(\Delta x)^2} + O[(\Delta x)^2] \qquad (1.54)$$

Analogous (and less accurate) formulae are valid within the NGF- and NGB-approximations, respectively.

For a consistent representation of *mixed* derivatives such as $f_{xy}$ one should use the same kind of approximation with respect to both the $x$- and the $y$-direction. (This may not hold if $x$ and $y$ have a different character, e.g. one space and one time variable; see Section 1.4.2 and Chapter 5.) In this way we find, using the Stirling expressions as an example,

$$
\begin{aligned}
[f_{xy}]_{i,j} &\approx \frac{1}{4\Delta x \Delta y}[f_{i+1,j+1} - f_{i+1,j-1} - f_{i-1,j+1} + f_{i-1,j-1}] + O[\Delta x \Delta y] \\
&\equiv \frac{\mu \delta_i}{\Delta x}[\frac{\mu \delta_j f_{i,j}}{\Delta y}] + O[\Delta x \Delta y]
\end{aligned}
\tag{1.55}
$$

The curvature of the function $f(x,y)$ at some point may be calculated by applying the nabla operator twice. There are two ways in which this operator $\nabla^2$ may be approximated. (Note that the nabla operator $\nabla$ mentioned in this paragraph should not be mixed up with the *backward difference* for which we use the same symbol.) Let us assume, just for simplicity of notation, that $\Delta y = \Delta x \equiv \Delta l$. Then we may either "difference" along the grid axes, writing the local curvature at the grid point $(i,j)$ as

$$
\nabla^2 f(x,y) \approx \frac{1}{(\Delta l)^2}[f_{i+1,j} + f_{i,j+1} + f_{i-1,j} + f_{i,j-1} - 4f_{i,j}]
\tag{1.56}
$$

or we may prefer to apply "diagonal differencing", writing

$$
\nabla^2 f(x,y) \approx \frac{1}{2(\Delta l)^2}[f_{i+1,j+1} + f_{i-1,j+1} + f_{i-1,j-1} + f_{i+1,j-1} - 4f_{i,j}]
\tag{1.57}
$$

## 1.4    Sample Applications

The wealth of applications of the finite difference formalism will become accessible only after a detailed consideration of linear algebra (Chapter 2) and of the ordinary and partial differential equations of physics (Chapters 4 and 5). Here we have to be content with a few hints which hopefully will whet the reader's appetite.

### 1.4.1    Classical Point Mechanics

The equations of motion of mass points in classical mechanics are ordinary differential equations of second order. Thus the physicist's favorite pet, the harmonic oscillator, obeys the equation of motion

$$
\frac{d^2 x}{dt^2} = -\omega_0^2 x
\tag{1.58}
$$

Everybody knows how to solve this equation analytically. What, then, is the procedure to follow in computational physics? We may, for once, replace the second differential quotient by the second Stirling-type difference quotient (see equ. 1.48):

$$\frac{\delta^2 x_k}{(\Delta t)^2} = -\omega_0^2 x_k + O[(\Delta t)^2]. \tag{1.59}$$

Assume now that the table of trajectory points, $\{x_k; \ k = 1, 2, \ldots\}$, be already known up to time $t_n$, and that we want to compute the next value $x_{n+1}$. From 1.59 we get

$$\frac{1}{(\Delta t)^2}(x_{n+1} - 2x_n + x_{n-1}) = -\omega_0^2 x_n + O[(\Delta t)^2] \tag{1.60}$$

or, explicitly,

$$x_{n+1} = 2x_n - x_{n-1} + (-\omega_0^2 x_n)(\Delta t)^2 + O[(\Delta t)^4] \tag{1.61}$$

In the field of statistical-mechanical simulation this formula is known as the Verlet algorithm [VESELY 78]. Of course, we may employ it also if 1.58 contains, instead of the harmonic acceleration term $-\omega_0^2 x$, any other continuous function of $x$. Anyone who has ever attempted to tackle by analytical means even the most simple of all anharmonic oscillators,

$$\frac{d^2 x}{dt^2} = -\omega_0^2\, x - \beta\, x^3 \tag{1.62}$$

will certainly appreciate this.

EXERCISE: a) Write a program to tabulate and/or display graphically the analytical solution to equ. 1.58. (You may achieve a very concise visualization by displaying the trajectory in *phase space*, i.e. in the coordinate system $\{x; \dot{x}\}$; where for $\dot{x}$ the approximation $\dot{x} \approx (x_{n+1} - x_{n-1})/2\Delta t$ may be used.) Choose specific values of $\omega_0^2$, $\Delta t$ and $x_0, \dot{x}_0$, and use these to determine the exact value of $x_1$. Then, starting with $x_0$ and $x_1$, employ the algorithm 1.61 to compute the further path $\{x_n \,; n = 2, 3, \ldots\}$. Test the performance of your program by varying $\Delta t$ and $\omega_0^2$.
b) Now apply your code to the *anharmonic* oscillator 1.62. To start the algorithm you may either use the exact value of $x_1$ (see, e.g., [LANDAU 62], Chap. V, §28), or the approximate value given by

$$x_1 \approx x_0 + \dot{x}_0 \Delta t + \ddot{x}_0 \frac{(\Delta t)^2}{2} \tag{1.63}$$

## 1.4.2    Diffusion and Thermal Conduction

The diffusion equation reads, in one dimension,

$$\frac{\partial u(x, t)}{\partial t} = D\frac{\partial^2 u(x, t)}{\partial x^2} \tag{1.64}$$

The variables $x$ and $t$ are again assumed to be discrete. Writing the desired density function $u$ at position $x_i$ at time $t_n$ as

$$u_i^n \equiv u(x_i, t_n) \, , \tag{1.65}$$

we may replace the time derivative $\partial u/\partial t$ by the linear DNGF-approximation (see equ. 1.32). For the second derivative by $x$ on the right hand side of 1.64 we use the Stirling approximation DDST (equ. 1.48) and obtain the so-called "FTCS scheme" (meaning "forward-time, centered-space"),

$$\frac{1}{\Delta t}[u_i^{n+1} - u_i^n] = \frac{D}{(\Delta x)^2}[u_{i+1}^n - 2u_i^n + u_{i-1}^n] \tag{1.66}$$

which will be considered in more detail in Section 5.2.1. Introducing the abbreviation $a \equiv D \, \Delta t/(\Delta x)^2$ we may rewrite this as an explicit formula,

$$u_i^{n+1} = (1 - 2a)u_i^n + a(u_{i-1}^n + u_{i+1}^n) \, , \tag{1.67}$$

which is valid for $i = 1, \ldots N - 1$. If the values of the function $u$ at the boundary points $x_0$ und $x_N$ are held fixed, and some initial values $u_i^0$ , $i = 0, \ldots N$ are assumed, the expression 1.67 determines the space-time evolution of $u$ uniquely.

EXERCISE: If we interpret $u(x,t)$ as an energy density, or simply as the temperature $T$, along a rod of length $L = 1$, equ. 1.64 may be understood as describing the conduction of heat, i.e. the spatio-temporal development of $T(x,t)$:

$$\frac{\partial T(x,t)}{\partial t} = \lambda \frac{\partial^2 T(x,t)}{\partial x^2} \tag{1.68}$$

Let us now divide the rod into 10 pieces of equal length, and assume the boundary conditions $T(0,t) \equiv T_0^n = 1.0$ and $T(L,t) \equiv T_{10}^n = 0.5$. The values for the temperature at time $t = 0$ (the *initial values*) are $T_1^0 = T_2^0 = \ldots T_{10}^0 = 0.5$ and $T_0^0 = 1.0$.

Employ equ. 1.67 to compute the distribution of temperatures at successive time steps; choose various values of the quantity $a$ (say, between 0.1 and 0.6. (See also the stability considerations in Section 5.2.1.)

# Chapter 2

# Linear Algebra

*Carl Friedrich Gauss eliminated unwanted matrix elements*

By the introduction of finite differences a function $f(x)$ depending on a single variable is converted into a table of function values. Such a table may be interpreted as a vector $\mathbf{f} \equiv (f_k; \ k = 1, \ldots, M)$. Similarly, a function of *two* variables may be tabulated in the format of a matrix:

$$\mathbf{F} \equiv [f_{i,j}] \equiv [f(x_i, y_j); \ i = 1, \ldots M; \ j = 1, \ldots N]. \qquad (2.1)$$

In many physical applications position and time are the relevant independent variables; in such cases the time variable $t$ will take the place of $y$. In particular, this holds whenever we have an *equation of motion* describing the temporal evolution of the quantity $f(x, t)$, i.e. a partial differential equation involving the derivatives of $f$ with respect to both independent variables. *Initial value problems* of this kind, when treated by the finite difference formalism, lead to systems of linear equations whose matrix has a specific, rather simple structure.

In contrast, in the case of stationary *boundary value problems* the variables $x$ and $y$ (and maybe a third independent variable $z$) are indeed spatial coordinates;

21

but again we have to do with partial differential equations which, by "differencing", may be transformed into systems of linear equations (see, e.g., equ. 5.84).

Further applications of linear algebra can be found in stochastics, where covariance matrices have to be handled (see Chapter 3,) and in quantum mechanics (eigenvalue problems.)

The fundamental manipulations we will have to perform on matrices are

- Inversion of a matrix:

$$\mathbf{A} \Longleftrightarrow \mathbf{A}^{-1} \tag{2.2}$$

- Finding the solution to the system of equations defined by a matrix $\mathbf{A}$ and a vector $\mathbf{b}$:

$$\mathbf{A} \cdot \mathbf{x} = \mathbf{b} \tag{2.3}$$

(To achieve this it is not necessary to determine the inverse $\mathbf{A}^{-1}$.)

- Finding the eigenvalues $\lambda_i$ and the eigenvectors $\mathbf{a}_i$ of a quadratic matrix:

$$\left. \begin{array}{rcl} |\mathbf{A} - \lambda_i \mathbf{I}| & = & 0 \\ (\mathbf{A} - \lambda_i \mathbf{I}) \cdot \mathbf{a}_i & = & 0 \end{array} \right\} \; i = 1, \ldots N \tag{2.4}$$

(Here, $|\mathbf{M}|$ denotes the determinant of a matrix.)

There are many excellent textbooks explaining the standard methods to employ for these tasks. And every computer center offers various subroutine libraries that contain well-proven tools for most problems one may encounter. In what follows we will only

- explain the standard techniques of linear algebra to such an extent as to render the above-mentioned *black box* subroutines at least semitransparent;

- explicate specific methods for the treatment of matrices which are either diagonally dominated or symmetric (or both).

## 2.1   Exact Methods

### 2.1.1   Gauss Elimination and Back Substitution

This is the classic technique for finding the solution of a system of linear algebraic equations $\mathbf{A} \cdot \mathbf{x} = \mathbf{b}$, with the special bonus of yielding the inverse $\mathbf{A}^{-1}$ as well. Let us write the given system of equations in the form

$$\begin{pmatrix} a_{11} & a_{12} & \cdot & \cdot & \cdot \\ a_{21} & a_{22} & & & \\ \cdot & & \cdot & & \\ \cdot & & & \cdot & \\ \cdot & & & & a_{NN} \end{pmatrix} \cdot \begin{pmatrix} x_1 \\ \cdot \\ \cdot \\ \cdot \\ x_N \end{pmatrix} = \begin{pmatrix} b_1 \\ \cdot \\ \cdot \\ \cdot \\ b_N \end{pmatrix} \tag{2.5}$$

If we could transform these equations in such a way that the matrix on the left-hand side were *triangular*, i.e.

$$
\begin{pmatrix}
a'_{11} & a'_{12} & \cdot & \cdot & \cdot \\
0 & a'_{22} & & & \\
\cdot & & \cdot & & \\
\cdot & & & \cdot & \\
0 & \cdot & \cdot & 0 & a'_{NN}
\end{pmatrix}
\cdot
\begin{pmatrix}
x_1 \\
\cdot \\
\cdot \\
\cdot \\
x_N
\end{pmatrix}
=
\begin{pmatrix}
b'_1 \\
\cdot \\
\cdot \\
\cdot \\
b'_N
\end{pmatrix}
\tag{2.6}
$$

this would all but solve the problem. In order to obtain this triangular form we use the following theorem:

> The solution vector $\mathbf{x}$ remains unchanged if arbitrary pairs of rows in the matrix $\mathbf{A}$ and in the vector $\mathbf{b}$ are interchanged *simultaneously*; more generally, replacing a row by a linear combination of itself and other rows leaves $\mathbf{x}$ unaltered.

This leads us to the following procedure:

---

**Gauss elimination:**

- Find the largest (by absolute value) element in the first column, and let $i$ be the row number of that element; exchange the first and the $i$-th row in $\mathbf{A}$ and $\mathbf{b}$.

- Subtract from the 2nd to $N$-th rows in $\mathbf{A}$ and $\mathbf{b}$ such multiples of the first row that all $a_{i1} = 0$.

- Repeat this procedure for the second column and row, etc., up to $N - 1$.

---

This method is called Gauss(ian) elimination with simple (partial) pivoting. In the more efficient method of complete pivoting not only rows but also columns are interchanged; this, however, involves memorizing all previous interchanges and is therefore more difficult to program.

Having transformed the matrix to the triangular form 2.6, we may now determine the elements of the solution vector by *back substitution* as described in the following box.

**Back substitution:**

$$x_N = \frac{1}{a'_{NN}} b'_N \qquad (2.7)$$

$$x_{N-1} = \frac{1}{a'_{N-1,N-1}} \left( b'_{N-1} - a'_{N-1,N}\, x_N \right) \qquad (2.8)$$

$$\vdots$$

$$x_i = \frac{1}{a'_{ii}} \left( b'_i - \sum_{j=i+1}^{N} a'_{ij}\, x_j \right) ; \quad i = N-2, \ldots, 1 \qquad (2.9)$$

If we need, in addition to the solution of our system of equations, also the *inverse* of the matrix $\mathbf{A}$, we simply have to apply the foregoing recipe *simultaneously* to $N$ unit vectors $\mathbf{b}_j = \mathbf{e}_j$ of the form

$$\mathbf{e}_1 \equiv \begin{pmatrix} 1 \\ 0 \\ . \\ . \\ 0 \end{pmatrix}, \ \mathbf{e}_2 \equiv \begin{pmatrix} 0 \\ 1 \\ 0 \\ . \end{pmatrix} \ \text{etc.} \qquad (2.10)$$

Following the triangulation of $\mathbf{A}$ we have $N$ new vectors $\mathbf{b}_j{}'$. Each of these is successively used in back substitution; each solution vector so obtained is then a *column vector* of the desired matrix $\mathbf{A}^{-1}$.

EXAMPLE: To determine the inverse of

$$\mathbf{A} = \begin{pmatrix} 3 & 1 \\ 2 & 4 \end{pmatrix}$$

we write

$$\begin{pmatrix} 3 & 1 \\ 2 & 4 \end{pmatrix} \cdot \begin{pmatrix} \alpha_{11} & \alpha_{12} \\ \alpha_{21} & \alpha_{22} \end{pmatrix} = \begin{pmatrix} 1 & 0 \\ 0 & 1 \end{pmatrix}$$

By (trivial) Gauss elimination we obtain the triangular system

$$\begin{pmatrix} 3 & 1 \\ 0 & \frac{10}{3} \end{pmatrix} \cdot \begin{pmatrix} \alpha_{11} & \alpha_{12} \\ \alpha_{21} & \alpha_{22} \end{pmatrix} = \begin{pmatrix} 1 & 0 \\ -\frac{2}{3} & 1 \end{pmatrix}$$

Back substitution in

$$\begin{pmatrix} 3 & 1 \\ 0 & \frac{10}{3} \end{pmatrix} \cdot \begin{pmatrix} \alpha_{11} \\ \alpha_{21} \end{pmatrix} = \begin{pmatrix} 1 \\ -\frac{2}{3} \end{pmatrix}$$

yields

$$\begin{pmatrix} \alpha_{11} \\ \alpha_{21} \end{pmatrix} = \begin{pmatrix} \frac{2}{5} \\ -\frac{1}{5} \end{pmatrix}$$

and from

$$\begin{pmatrix} 3 & 1 \\ 0 & \frac{10}{3} \end{pmatrix} \cdot \begin{pmatrix} \alpha_{12} \\ \alpha_{22} \end{pmatrix} = \begin{pmatrix} 0 \\ 1 \end{pmatrix}$$

we find

$$\begin{pmatrix} \alpha_{12} \\ \alpha_{22} \end{pmatrix} = \begin{pmatrix} -\frac{1}{10} \\ \frac{3}{10} \end{pmatrix}$$

so that

$$\mathbf{A}^{-1} = \begin{pmatrix} \frac{2}{5} & -\frac{1}{10} \\ -\frac{1}{5} & \frac{3}{10} \end{pmatrix}$$

## 2.1.2   Simplifying Matrices: The Householder Transformation

Gaussian elimination may be understood as a transformation of the given matrix to triangular form. Without saying so we have successively applied the following transformation to the given system of equations, $\mathbf{A} \cdot \mathbf{x} = \mathbf{b}$:

$$\mathbf{P}_{n-1} \ldots \mathbf{P}_1 \cdot \mathbf{A} \cdot \mathbf{x} = \mathbf{P}_{n-1} \ldots \mathbf{P}_1 \cdot \mathbf{b} \qquad (2.11)$$

with the requirement that the transformed matrix $\mathbf{P}_{n-1} \ldots \mathbf{P}_1 \cdot \mathbf{A}$ be triangular. The Gaussian recipe, resulting in a specific sequence of matrices $\mathbf{P}_i$, is not unique in achieving triangularity. Many other choices of transformation matrices are possible, and the respective techniques are labeled by the names of their authors, such as Givens, Schmidt, or Householder.

More generally, in applications of linear algebra the first step is often the transformation of a given matrix to some simple form. This may be the triangular form as above, or a tri-diagonal structure as we will discuss later on. In all such cases we wish to strip certain column and/or row vectors of some of their rear elements.

The basic operation in such repeated truncation procedures is known as the *Householder transformation*.

Let $\mathbf{a}$ be a vector (such as a matrix column vector), and $\mathbf{e}_1$ a unit vector as in equ. 2.10. We define an auxiliary vector $\mathbf{b} \equiv \mathbf{a} \pm |\mathbf{a}| \, \mathbf{e}_1$ and normalize it as $\mathbf{b}_0 \equiv \mathbf{b}/|\mathbf{b}|$. (Either the plus or the minus sign may be used; one or the other may be better in terms of numerical accuracy, depending on the elements of $\mathbf{A}$.) The *Householder matrix* $\mathbf{P}$ is then defined by

$$\mathbf{P} \equiv \mathbf{1} - 2\,\mathbf{b}_0 \cdot \mathbf{b}_0^T \qquad (2.12)$$

It is easy to see that in the transformed vector $\mathbf{a}' \equiv \mathbf{P} \cdot \mathbf{a}$ all elements but the first are equal to zero.

This transformation may be applied successively to the first, second, etc. column/row vectors of a given matrix, thus eliminating sub- or off-diagonal elements as desired. Technical details of these applications may be found in [PRESS 86, WILKINSON 67].

EXAMPLE: The following method to triangularize a $3 \times 3$ matrix is pathetically inefficient. It is discussed here only to demonstrate the principle of the Householder transformation that is such an ubiquitous ingredient in linear algebra black box routines.
Starting out with

$$\mathbf{A} = \begin{pmatrix} 1 & 2 & 3 \\ 4 & 5 & 6 \\ 7 & 8 & 9 \end{pmatrix} \tag{2.13}$$

we pick the first column vector $\mathbf{a} \equiv (1,4,7)^T$ to construct the auxiliary vector (using the minus sign in the above definition) $\mathbf{b} = (-7.124, 4, 7)^T$ which is then normalized to $\mathbf{b}_0 = (-0.662, 0.372, 0.651)^T$. The resulting Householder matrix is

$$\mathbf{P}_1 = \begin{pmatrix} 0.123 & 0.492 & 0.862 \\ 0.492 & 0.724 & -0.484 \\ 0.862 & -0.484 & 0.153 \end{pmatrix} \tag{2.14}$$

And indeed, when we multiply $\mathbf{A}$ by $\mathbf{P}_1$ the resulting matrix has a stripped first column vector:

$$\mathbf{P}_1 \cdot \mathbf{A} = \begin{pmatrix} 8.124 & 2.708 & 11.078 \\ 0 & 4.602 & 1.464 \\ 0 & -0.696 & 1.062 \end{pmatrix} \tag{2.15}$$

Next we concentrate on the lower right $2 \times 2$ submatrix of $\mathbf{P}_1 \cdot \mathbf{A}$. From its first column vector, $\mathbf{a} = (4.602, -0.696)^T$ we construct a $2 \times 2$ Householder matrix which we then promote to a $3 \times 3$ matrix by adding a trivial first line and column, respectively:

$$\mathbf{P}_2 = \begin{pmatrix} 1 & 0 & 0 \\ 0 & 0.989 & -0.149 \\ 0 & -0.149 & -0.989 \end{pmatrix} \tag{2.16}$$

Checking the total result, we find that the matrix

$$\mathbf{P}_2 \cdot \mathbf{P}_1 \cdot \mathbf{A} = \begin{pmatrix} 8.124 & 2.708 & 11.078 \\ 0 & 4.655 & 1.289 \\ 0 & 0 & -1.269 \end{pmatrix} \tag{2.17}$$

has the required triangular structure.

## 2.1.3   LU Decomposition

A more modern, and in some respects more efficient, device for the solution of a linear system than Gauss elimination is due to the authors Banachiewicz, Cholesky and Crout. The name "LU decomposition" implies a "lower-upper" factorization of the given matrix. In other words, we seek to represent the matrix $\mathbf{A}$ as a product of two triangular matrices, such that

$$\mathbf{A} = \mathbf{L} \cdot \mathbf{U} \tag{2.18}$$

with

$$\mathbf{L} = \begin{pmatrix} l_{11} & 0 & . & 0 \\ l_{21} & l_{22} & . & . \\ . & & . & 0 \\ l_{N1} & . & . & l_{NN} \end{pmatrix} ; \ \mathbf{U} = \begin{pmatrix} u_{11} & u_{12} & . & u_{1N} \\ 0 & u_{22} & . & . \\ . & & . & . \\ 0 & . & 0 & u_{NN} \end{pmatrix} \tag{2.19}$$

Writing $\mathbf{A} \cdot \mathbf{x} = \mathbf{b}$ as

$$\mathbf{L} \cdot (\mathbf{U} \cdot \mathbf{x}) = \mathbf{b} \tag{2.20}$$

we can split up the task according to

$$\mathbf{L} \cdot \mathbf{y} = \mathbf{b} \tag{2.21}$$

and

$$\mathbf{U} \cdot \mathbf{x} = \mathbf{y} \tag{2.22}$$

Owing to the triangular form of the matrices $\mathbf{L}$ and $\mathbf{U}$ these equations are easy to solve. First we compute an auxiliary vector $\mathbf{y}$ by *forward substitution*:

$$y_1 = \frac{1}{l_{11}} b_1 \tag{2.23}$$

$$y_i = \frac{1}{l_{ii}} \left( b_i - \sum_{j=1}^{i-1} l_{ij} y_j \right) ; \quad i = 2, \dots, N \tag{2.24}$$

The solution vector $\mathbf{x}$ is then obtained by *back substitution* in the same manner as in the Gauss elimination technique:

$$x_N = \frac{1}{u_{NN}} y_N \tag{2.25}$$

$$x_i = \frac{1}{u_{ii}} \left( y_i - \sum_{j=i+1}^{N} u_{ij} x_j \right) ; \quad i = N-1, \dots, 1 \tag{2.26}$$

How, then, are we to find the matrices $\mathbf{L}$ and $\mathbf{U}$? The definition $\mathbf{L} \cdot \mathbf{U} = \mathbf{A}$ is equivalent to the $N^2$ equations

$$\sum_{k=1}^{N} l_{ik} u_{kj} = a_{ij} ; \ i = 1, \dots N; \ j = 1, \dots N \tag{2.27}$$

We are free to choose $N$ out of the $N^2 + N$ unknowns $l_{ij}, u_{ij}$. For convenience, we put $l_{ii} = 1 \, (i = 1, \dots N)$. Also, due to the triangular structure of $\mathbf{L}$ and $\mathbf{U}$, the summation index $k$ will not run over the whole interval $[1, \ldots, N]$. Rather, we have

$$\text{for } i \leq j : \quad \sum_{k=1}^{i} l_{ik} u_{kj} = a_{ij} \tag{2.28}$$

$$\text{for } i > j : \quad \sum_{k=1}^{j} l_{ik} u_{kj} = a_{ij} \tag{2.29}$$

This leads to the following procedure for the evaluation of $u_{ij}$ and $l_{ij}$, as given by Crout:

---

**LU decomposition:** For $j = 1, 2, \ldots N$ compute

$$u_{1j} = a_{1j} \tag{2.30}$$

$$u_{ij} = a_{ij} - \sum_{k=1}^{i-1} l_{ik} u_{kj}; \quad i = 2, \ldots, j \tag{2.31}$$

$$l_{ij} = \frac{1}{u_{jj}} \left( a_{ij} - \sum_{k=1}^{j-1} l_{ik} u_{kj} \right); \quad i = j+1, \ldots, N \tag{2.32}$$

---

The *determinant* of the given matrix is obtained as a side result of LU decomposition:

$$|\mathbf{A}| = |\mathbf{L}| \cdot |\mathbf{U}| = u_{11} u_{22} \ldots u_{NN} \tag{2.33}$$

EXAMPLE: For the LU decomposition of

$$\mathbf{A} = \begin{pmatrix} 1 & 2 \\ 3 & 4 \end{pmatrix}$$

we find, according to Crout:

$$\begin{aligned}
j = 1, \ i = 1: \quad u_{11} &= a_{11} = 1 \\
j = 1, \ i = 2: \quad l_{21} &= \frac{1}{u_{11}} a_{21} = 3
\end{aligned}$$

$$\begin{aligned}
j = 2, \ i = 1: \quad u_{12} &= a_{12} = 2 \\
j = 2, \ i = 2: \quad u_{22} &= a_{22} - l_{21} u_{12} = -2
\end{aligned}$$

so that

$$\begin{pmatrix} 1 & 2 \\ 3 & 4 \end{pmatrix} = \underbrace{\begin{pmatrix} 1 & 0 \\ 3 & 1 \end{pmatrix}}_{\mathbf{L}} \cdot \underbrace{\begin{pmatrix} 1 & 2 \\ 0 & -2 \end{pmatrix}}_{\mathbf{U}}$$

At each step $(j, i)$ the required elements $l_{ik}$, $u_{kj}$ are already available. Each of the elements $a_{ij}$ of the original matrix $\mathbf{A}$ is used only once. In a computer code one may therefore save storage space by overwriting $a_{ij}$ by $u_{ij}$ or $l_{ij}$, respectively. (The $l_{ii}$ are equal to 1 and need not be stored at all.)

Speaking of computer codes: the above procedure is only the basic principle of the LU decomposition technique. In order to write an efficient program one would have to include pivoting, which is more involved here than in the Gaussian elimination method (see [PRESS 86], p.34f.).

An important advantage of LU decomposition as compared to Gauss' method is the fact that the vector $\mathbf{b}$ has so far not been manipulated at all. (In particular, there was no exchanging of rows etc.) Only for the calculation of a solution vector $\mathbf{x}$ by forward and backward substitution the elements of $\mathbf{b}$ come into play. In other words, we may use the factors $\mathbf{L}$ and $\mathbf{U}$ of a given matrix $\mathbf{A}$ again and again, with different vectors $\mathbf{b}$.

If required, the *inverse* of the matrix $\mathbf{A}$ may again be determined in the same manner as with Gaussian elimination: after solving the equations $\mathbf{A} \cdot \mathbf{x}_j = \mathbf{e}_j$, with the $N$ unit vectors $\mathbf{e}_j$, one combines the column vectors $\mathbf{x}_j$ to find $\mathbf{A}^{-1}$.

## 2.1.4    Tridiagonal Matrices: Recursion Method

In many applications the matrix $\mathbf{A}$ in the system of equations $\mathbf{A} \cdot \mathbf{x} = \mathbf{b}$ has non-zero elements only along the main diagonal and in the immediately adjacent diagonals. Or we may have applied the Householder transformation to a given matrix such that a tridiagonal structure results. In all such cases a very fast method may be used to find the solution vector $\mathbf{x}$. With the notation

$$\mathbf{A} \equiv \begin{pmatrix} \beta_1 & \gamma_1 & 0 & . & . & 0 \\ \alpha_2 & \beta_2 & \gamma_2 & 0 & . & 0 \\ 0 & \alpha_3 & \beta_3 & \gamma_3 & 0 & . \\ . & . & . & . & . & . \\ . & . & . & \alpha_{N-1} & \beta_{N-1} & \gamma_{N-1} \\ . & . & . & 0 & \alpha_N & \beta_N \end{pmatrix} \tag{2.34}$$

the system of equations reads

$$\begin{aligned} \beta_1 x_1 + \gamma_1 x_2 &= b_1 \\ \alpha_i x_{i-1} + \beta_i x_i + \gamma_i x_{i+1} &= b_i ; \quad i = 2, \ldots, N-1 \\ \alpha_N x_{N-1} + \beta_N x_N &= b_N \end{aligned} \tag{2.35}$$

Introducing auxiliary variables $g_i$ and $h_i$ by the recursive ansatz

$$x_{i+1} = g_i x_i + h_i ; \quad i = 1, \ldots, N-1 \tag{2.36}$$

we find from 2.35 the "downward recursion formulae"

$$g_{N-1} = \frac{-\alpha_N}{\beta_N} \quad , \quad h_{N-1} = \frac{b_N}{\beta_N} \tag{2.37}$$

$$g_{i-1} = \frac{-\alpha_i}{\beta_i + \gamma_i\, g_i} \quad , \quad h_{i-1} = \frac{b_i - \gamma_i\, h_i}{\beta_i + \gamma_i\, g_i}; \quad i = N-1, \ldots, 2 \tag{2.38}$$

Having arrived at $g_1$ and $h_1$ we insert the known values of $g_i$, $h_i$ in the "upward recursion formulae"

$$x_1 = \frac{b_1 - \gamma_1\, h_1}{\beta_1 + \gamma_1\, g_1} \tag{2.39}$$

$$x_{i+1} = g_i\, x_i + h_i; \quad i = 1, \ldots, N-1 \tag{2.40}$$

(Equation 2.39 for the starting value $x_1$ follows from $\beta_1 x_1 + \gamma_1 x_2 = b_1$ and $x_2 = g_1 x_1 + h_1$.)

EXAMPLE: In $\mathbf{A} \cdot \mathbf{x} = \mathbf{b}$, let

$$\mathbf{A} \equiv \begin{pmatrix} \beta_1 & \gamma_1 & 0 & 0 \\ \alpha_2 & \beta_2 & \gamma_2 & 0 \\ 0 & \alpha_3 & \beta_3 & \gamma_3 \\ 0 & 0 & \alpha_4 & \beta_4 \end{pmatrix} = \begin{pmatrix} 2 & 1 & 0 & 0 \\ 2 & 3 & 1 & 0 \\ 0 & 1 & 4 & 2 \\ 0 & 0 & 1 & 3 \end{pmatrix} \quad \text{and} \quad \mathbf{b} = \begin{pmatrix} 1 \\ 2 \\ 3 \\ 4 \end{pmatrix}$$

Downward recursion (Equ. 2.37, 2.38):

$$g_3 = -\frac{1}{3} \quad , \quad h_3 = \frac{4}{3}$$

$$i = 3: \quad g_2 = -\frac{3}{10} \quad , \quad h_2 = \frac{1}{10}$$

$$i = 2: \quad g_1 = -\frac{20}{27} \quad , \quad h_1 = \frac{19}{27}$$

Upward recursion (Equ. 2.39, 2.40):

$$x_1 = \frac{8}{34}$$

$$i = 1: \quad x_2 = \frac{9}{17}$$

$$i = 2: \quad x_3 = -\frac{1}{17}$$

$$i = 3: \quad x_4 \;=\; \frac{23}{17}$$

A similar method which may be used in the case of a five-diagonal matrix is given in [ENGELN 91].

## 2.2 Iterative Methods

The methods described so far for the solution of linear systems are – in principle – exact. Any numerical errors are due to the finite machine accuracy (see Appendix A). If the given matrices are well-behaved, the process of *pivoting* explained earlier keeps those roundoff errors small. However, if the matrices are near singular, errors may be amplified in an inconvenient way in the course of determining the solution. In such cases one should "cleanse" the solution by a method called *iterative improvement*.

Let $\mathbf{x}$ be the exact solution of $\mathbf{A} \cdot \mathbf{x} = \mathbf{b}$, and let $\mathbf{x}'$ be a still somewhat inaccurate (or simply estimated) solution vector, such that

$$\mathbf{x} \equiv \mathbf{x}' + \delta\,\mathbf{x} \tag{2.41}$$

Inserting this into the given equation we find

$$\boxed{\mathbf{A} \cdot \delta\mathbf{x} = \mathbf{b} - \mathbf{A} \cdot \mathbf{x}'} \tag{2.42}$$

Since the right-hand side of this equation contains known quantities only, we can use it to calculate $\delta\,\mathbf{x}$ and therefore $\mathbf{x}$. The numerical values in $\mathbf{b} - \mathbf{A} \cdot \mathbf{x}'$ are small, and double precision should be used here. If the LU decomposition of the matrix $\mathbf{A}$ is known, $\delta\,\mathbf{x}$ is most suitably found by forward and back substitution; only $\approx N^2$ operations are required in this case. In contrast, the "exact" methods we may have used to find $\mathbf{x}'$ take some $N^3$ operations.

EXAMPLE: The principle of iterative improvement may be demonstrated using a grossly inaccurate first approximation $\mathbf{x}'$. Let

$$\mathbf{A} = \begin{pmatrix} 1 & 2 \\ 3 & 4 \end{pmatrix}, \; \mathbf{b} = \begin{pmatrix} 3 \\ 2 \end{pmatrix} \; \text{and} \; \mathbf{x}' = \begin{pmatrix} -3 \\ 4 \end{pmatrix}$$

From

$$\mathbf{A} \cdot \delta\mathbf{x} = \begin{pmatrix} 3 \\ 2 \end{pmatrix} - \begin{pmatrix} 1 & 2 \\ 3 & 4 \end{pmatrix} \cdot \begin{pmatrix} -3 \\ 4 \end{pmatrix} = \begin{pmatrix} -2 \\ -5 \end{pmatrix}$$

we find, using the decomposition

$$\mathbf{L} = \begin{pmatrix} 1 & 0 \\ 3 & 1 \end{pmatrix} \; \text{and} \; \mathbf{U} = \begin{pmatrix} 1 & 2 \\ 0 & -2 \end{pmatrix}$$

the correction vector

$$\delta\mathbf{x} = \begin{pmatrix} -1 \\ -\frac{1}{2} \end{pmatrix}$$

so that the correct solution

$$\mathbf{x} = \begin{pmatrix} -4 \\ \frac{7}{2} \end{pmatrix}$$

is obtained.

The idea underlying the technique of iterative improvement may be extended in a very fruitful way. Let us interpret equ. 2.42 as an iterative formula,

$$\mathbf{A} \cdot (\mathbf{x}_{k+1} - \mathbf{x}_k) = \mathbf{b} - \mathbf{A} \cdot \mathbf{x}_k \qquad (2.43)$$

forgoing the ambition to reach the correct answer in one single step. We may then replace $\mathbf{A}$ on the left hand side by a matrix $\mathbf{B}$ which should not be too different from $\mathbf{A}$, but may be easier to invert:

$$\mathbf{B} \cdot (\mathbf{x}_{k+1} - \mathbf{x}_k) = \mathbf{b} - \mathbf{A} \cdot \mathbf{x}_k \qquad (2.44)$$

or

$$\mathbf{x}_{k+1} = \mathbf{B}^{-1} \cdot \mathbf{b} + \mathbf{B}^{-1} \cdot [\mathbf{B} - \mathbf{A}] \cdot \mathbf{x}_k \qquad (2.45)$$

This procedure can be shown to converge to the solution of $\mathbf{A} \cdot \mathbf{x} = \mathbf{b}$ if, and only if, $|\mathbf{x}_{k+1} - \mathbf{x}_k| < |\mathbf{x}_k - \mathbf{x}_{k-1}|$. This, however, is the case if all eigenvalues of the matrix

$$\mathbf{B}^{-1} \cdot [\mathbf{B} - \mathbf{A}]$$

are situated within the unit circle.

It is the choice of the matrix $\mathbf{B}$ where the various iterative methods differ. The three most important methods are known as *Jacobi relaxation*, *Gauss-Seidel relaxation* (GSR) and *successive over-relaxation* (SOR). In each of these techniques only such matrix manipulations occur that need less than $\approx N^3$ operations per iteration; usually $\approx N^2$ operations are necessary. For large matrices iterative methods are therefore much faster than the exact techniques.

## 2.2.1   Jacobi Relaxation

We first divide the given matrix according to

$$\mathbf{A} = \mathbf{D} + \mathbf{L} + \mathbf{R} \qquad (2.46)$$

where $\mathbf{D}$ contains only the diagonal elements of $\mathbf{A}$, while $\mathbf{L}$ and $\mathbf{R}$ are the left and right parts of $\mathbf{A}$, respectively. (The matrix $\mathbf{L}$ introduced here has, of course, nothing to do with the one defined earlier, in the framework of LU factorization).

The condition of being easy to invert is most readily met by the diagonal matrix **D**. We therefore choose $\mathbf{B} = \mathbf{D}$ and write the iteration formula 2.45 as

$$\mathbf{D} \cdot \mathbf{x}_{k+1} = \mathbf{b} + [\mathbf{D} - \mathbf{A}] \cdot \mathbf{x}_k \qquad (2.47)$$

or

$$a_{ii}\, x_i^{(k+1)} = b_i - \sum_{j \neq i} a_{ij}\, x_j^{(k)}; \quad i = 1, \ldots, N \qquad (2.48)$$

EXAMPLE: In $\mathbf{A} \cdot \mathbf{x} = \mathbf{b}$ let

$$\mathbf{A} = \begin{pmatrix} 3 & 1 \\ 2 & 4 \end{pmatrix}; \quad \mathbf{b} = \begin{pmatrix} 3 \\ 2 \end{pmatrix}$$

Starting from the estimated solution

$$\mathbf{x}_0 = \begin{pmatrix} 1.2 \\ 0.2 \end{pmatrix}$$

and using the diagonal part of **A**,

$$\mathbf{D} = \begin{pmatrix} 3 & 0 \\ 0 & 4 \end{pmatrix}$$

in the iteration we find the increasingly more accurate solutions

$$\mathbf{x}_1 = \begin{pmatrix} 0.933 \\ -0.100 \end{pmatrix}; \quad \mathbf{x}_2 = \begin{pmatrix} 1.033 \\ 0.033 \end{pmatrix} \; etc. \; \rightarrow \mathbf{x}_\infty = \begin{pmatrix} 1 \\ 0 \end{pmatrix}$$

The Jacobi method converges best for diagonally dominated matrices **A**, but even there the rate of convergence is moderate at best. The convergence behavior is governed by the eigenvalues of the matrix $-[\mathbf{L} + \mathbf{R}]$. Writing the Jacobi scheme in the form

$$\mathbf{x}_{k+1} = \mathbf{D}^{-1} \cdot \mathbf{b} + \mathbf{J} \cdot \mathbf{x}_k, \qquad (2.49)$$

with the *Jacobi block matrix*

$$\mathbf{J} \equiv \mathbf{D}^{-1} \cdot [\mathbf{D} - \mathbf{A}] = -\mathbf{D}^{-1} \cdot [\mathbf{L} + \mathbf{R}] \qquad (2.50)$$

convergence requires that all eigenvalues of **J** be smaller than one (by absolute value). Denoting the largest eigenvalue (the *spectral radius*) of **J** by $\lambda_J$, we have for the asymptotic rate of convergence

$$r_J \equiv \frac{|\mathbf{x}_{k+1} - \mathbf{x}_k|}{|\mathbf{x}_k - \mathbf{x}|} \approx |\lambda_J - 1| \qquad (2.51)$$

In the above example $\lambda_J = 0.408$ and $r \approx 0.59$.

## 2.2.2   Gauss-Seidel Relaxation (GSR)

We obtain a somewhat faster convergence than in the Jacobi scheme if we choose $\mathbf{B} = \mathbf{D} + \mathbf{L}$, writing the iteration as

$$[\mathbf{D} + \mathbf{L}] \cdot \mathbf{x}_{k+1} = \mathbf{b} - \mathbf{R} \cdot \mathbf{x}_k \qquad (2.52)$$

Solving the set of *implicit* equations

$$a_{ii} x_i^{(k+1)} + \sum_{j<i} a_{ij} x_i^{(k+1)} = b_i - \sum_{j>i} a_{ij} x_j^{(k)} \, ; \; i = 1, \ldots, N \qquad (2.53)$$

is not quite as simple as solving the *explicit* Jacobi equations 2.48. However, since the matrix $\mathbf{D} + \mathbf{L}$ is triangular the additional effort is affordable.

EXAMPLE: With the same data as in the previous example we find the first two improved solutions

$$\mathbf{x}_1 = \begin{pmatrix} 0.933 \\ 0.033 \end{pmatrix} \, ; \; \mathbf{x}_2 = \begin{pmatrix} 0.989 \\ 0.006 \end{pmatrix}.$$

The convergence rate of the GSR scheme is governed by the matrix

$$\mathbf{G} \equiv - [\mathbf{D} + \mathbf{L}]^{-1} \cdot \mathbf{R} \qquad (2.54)$$

It can be shown [STOER 89] that the spectral radius of $\mathbf{G}$ is given by

$$\lambda_G = \lambda_J^2 \qquad (2.55)$$

so that the rate of convergence is now

$$r_G \approx \left| \lambda_J^2 - 1 \right| \qquad (2.56)$$

In our example $\lambda_G = 0.17$ and $r \approx 0.83$.

## 2.2.3   Successive Over-Relaxation (SOR)

This method, which is also called *simultaneous over-relaxation*, is based on the iteration ansatz

$$\mathbf{x}_{k+1}^{SOR} = \omega \, \mathbf{x}_{k+1}^{GSR} + (1 - \omega) \mathbf{x}_k \qquad (2.57)$$

The "relaxation parameter" $\omega$ may be varied within the range $0 \leq \omega \leq 2$ to optimize the method.

At each iteration step, then, the "old" vector $\mathbf{x}_k$ is mixed with the new vector $\mathbf{x}_{k+1}$ which has been calculated using GSR. Reshuffling equ. 2.57 we find

$$[\mathbf{D} + \mathbf{L}] \cdot \mathbf{x}_{k+1} = \omega \, \mathbf{b} - [\mathbf{R} - (1 - \omega) \, \mathbf{A}] \cdot \mathbf{x}_k \qquad (2.58)$$

A single row in this system of equations reads

$$a_{ii}\, x_i^{(k+1)} + \sum_{j<i} a_{ij}\, x_j^{(k+1)} \;=\; \omega\, b_i - \omega \sum_{j>i} a_{ij}\, x_j^{(k)} +$$

$$+(1-\omega) \sum_{j\leq i} a_{ij} x_j(k) \qquad i = 1, \ldots, N \qquad (2.59)$$

The rate of convergence of this procedure is governed by the matrix

$$\mathbf{S} \equiv -\left[\mathbf{D} + \mathbf{L}\right]^{-1} \cdot \left[\mathbf{R} - (1-\omega)\,\mathbf{A}\right] \qquad (2.60)$$

Again we may find a relation between the eigenvalues of $\mathbf{S}$ and those of $\mathbf{J}$: the optimal value of $\omega$ is given by [STOER 89]

$$\omega = \frac{2}{1 + \sqrt{1 - \lambda_J^2}} \qquad (2.61)$$

yielding

$$\lambda_S = \left[\frac{\lambda_J}{1 + \sqrt{1 - \lambda_J^2}}\right]^2 \qquad (2.62)$$

The asymptotic rate of convergence is

$$r_S \approx |\lambda_S - 1| \qquad (2.63)$$

EXAMPLE: With the same data as before we find from 2.61 an optimal relaxation parameter $\omega = 1.046$, and from that $r_s = 0.95$. The first two iterations yield

$$\mathbf{x}_1 = \begin{pmatrix} 0.921 \\ 0.026 \end{pmatrix}; \; \mathbf{x}_2 = \begin{pmatrix} 0.994 \\ 0.003 \end{pmatrix}.$$

The parameter $\omega$ as evaluated according to 2.61 is "optimal" only in the asymptotic sense, that is, after a certain number of iterations. During the first few iterative steps the SOR procedure may give rise to overshooting corrections – particularly if $\omega$ is distinctly larger than 1. One can avoid this delay of convergence by starting out with a value of $\omega = 1$, letting $\omega$ gradually approach the value given in 2.61. This procedure, which is known as "Chebysheff acceleration", consists of the following steps:

- The solution vector $\mathbf{x}$ is split in 2 vectors $\mathbf{x}_e$, $\mathbf{x}_o$ consisting of the elements $x_i$ with even and odd indices, respectively; the vector $\mathbf{b}$ is split up in the same manner.

- The two subvectors $\mathbf{x}_e$ and $\mathbf{x}_o$ are iterated in alternating succession, with the relaxation parameter being adjusted according to

$$
\begin{aligned}
\omega^{(0)} &= 1 \\
\omega^{(1)} &= \frac{1}{1 - \lambda_j^2/2} \\
\omega^{(k+1)} &= \frac{1}{1 - \lambda_j^2 \omega^{(k)}/4}, \quad k = 1, \ldots
\end{aligned}
\tag{2.64}
$$

## 2.2.4   Alternating Direction Implicit Method (ADI)

Chapter 5 will be devoted to the treatment of those partial differential equations which are of major importance in physics. In many cases the discretization of such PDEs yields systems of linear equations whose matrix is "almost" tridiagonal. More specifically, $\mathbf{A}$ has the following five-diagonal form:

$$
\mathbf{A} = \begin{pmatrix}
x & x &   &   & x &   &   & \\
x & x & x &   &   & x &   & \\
  & x & x & x &   &   & x & \\
  &   & x & x & x &   &   & x \\
x &   &   & x & x & x &   & \\
  & x &   &   & x & x & x & \\
  &   & x &   &   & x & x & \ddots \\
  &   &   & \ddots &   &   & \ddots & \ddots
\end{pmatrix}
\tag{2.65}
$$

In such cases it is feasible and advantageous to rewrite the system of equations in such a way that two coupled *tridiagonal* systems are obtained. This may be interpreted as treating the original system first row by row and then column by column. (There we have a partial explanation of the name *alternating direction implicit* method.) To achieve a consistent solution this procedure must be iterated, and once more a relaxation parameter is introduced and adjusted for optimum convergence.

The ADI scheme is tailored to the numerical treatment of the potential equation $\nabla^2 u = -\rho$. We therefore postpone a more detailed description of this method to Section 5.3.2. For the time being, suffice it to say that the method converges even more rapidly than SOR accelerated à la Chebysheff.

## 2.2.5   Conjugate Gradient Method (CG)

The task of solving the equation $\mathbf{A} \cdot \mathbf{x} = \mathbf{b}$ may be interpreted as a minimization problem. Defining the scalar function

$$
f(\mathbf{x}) \equiv \frac{1}{2} \, |\mathbf{A} \cdot \mathbf{x} - \mathbf{b}|^2 \,,
\tag{2.66}
$$

we only need to find that $N$-vector $\mathbf{x}$ which minimizes $f(\mathbf{x})$ (with the minimum value $f = 0$.)

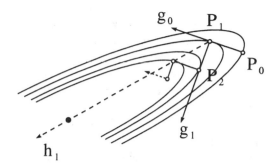

Figure 2.1: Conjugate gradients: $\mathbf{g}_0 = -\boldsymbol{\nabla} f(P_0)$ denotes the direction of steepest descent at point $P_0$, $\mathbf{g}_1$ is the same at point $P_1$, etc.; $\mathbf{h}_1$ points out the direction of the gradient conjugate to $\mathbf{g}_0$. The steepest descent method follows the tedious zig-zag course $P_0 \rightarrow P_1 \rightarrow P_2 \rightarrow \dots$ . The conjugate gradient $\mathbf{h}_1$ gets us to the goal in just two steps.

Various methods are available for the minimization of a scalar function of $N$ variables. In our case $f(\mathbf{x})$ is a quadratic function of $\mathbf{x}$, and in such instances the method of *conjugate gradients* is particularly efficient. There will be no matrix inversion at all – in marked contrast to the other iterative methods. However, the multiplication $\mathbf{A} \cdot \mathbf{x}$ must be performed several times, so that the procedure is economical only for sparse matrices $\mathbf{A}$. (For such matrices the multiplication will of course be done by specific subroutines involving less than $N^2$ operations.)

In order to explain the CG method we start out from the older and less efficient *steepest descent method* introduced by Cauchy. For simplicity of visualization, but without restriction of generality, we assume the function $f$ to depend on two variables $\mathbf{x} = (x_1, x_2)$ only. The lines of equal elevation of a quadratic function are ellipses that may, in adverse cases, have a very elongated shape, forming a long and narrow channel (see Fig. 2.1). Starting from some point $P_0$ with a position vector $\mathbf{x}_0$ and proceeding by *steepest descent* we would follow the local gradient

$$\mathbf{g}_0 = -\boldsymbol{\nabla} f(P_0) \tag{2.67}$$

As the figure shows (and as every alpine hiker knows) this direction will by no means lead directly to the extremal point of $f$. The best we can do – and this is indeed the next step in the steepest descent technique – is to proceed to the lowest point $P_1$ along the path that cuts through the narrow channel in the direction of $\mathbf{g}_0$. If we now determine once more the local gradient $\mathbf{g}_1 = -\boldsymbol{\nabla} f(P_1)$, it must be perpendicular (by construction) to $\mathbf{g}_0$. Iterating this procedure we arrive, after many mutually orthogonal bends, at the bottom of the channel.

We would arrive at our goal much faster if from point $P_1$ we took a path along the direction $\mathbf{h}_1$ instead of $\mathbf{g}_1$. But how are we to find $\mathbf{h}_1$? – Let us require that in proceeding along $\mathbf{h}_1$ the *change of the gradient of f* should have no component parallel to $\mathbf{g}_0$. (In contrast, when we follow $\mathbf{g}_1$, the *gradient of f* has – initially at least – no $\mathbf{g}_0$-component; this, however, changes very soon, and the lengthy zig-zag path ensues.) If we can achieve this, a gradient in the direction of $\mathbf{g}_0$ will not develop immediately – in fact, on quadratic surfaces it will never build up again. In our two-dimensional example this means that $\mathbf{h}_1$ must already point to the desired minimum.

If we apply these considerations to the particular quadratic function 2.66 we are led to the prescription given in Fig. 2.2.

If the system of equations – and therefore the surface $f(x_1, x_2)$ – is of dimension 2 only, we have reached our goal after the two steps described in Fig. 2.2, and $\mathbf{x} = \mathbf{x}_2$ is the solution vector. For systems of higher dimensionality one has to go on from $\mathbf{x}_2$ in the direction

$$\mathbf{h}_2 = \mathbf{g}_2 - \frac{\mathbf{g}_2 \cdot \mathbf{A} \cdot \mathbf{h}_1}{\mathbf{h}_1 \cdot \mathbf{A} \cdot \mathbf{h}_1} \mathbf{h}_1 \,. \tag{2.75}$$

until the next "low point" is reached at

$$\mathbf{x}_3 = \mathbf{x}_2 + \lambda_3 \, \mathbf{h}_2 \,, \tag{2.76}$$

with

$$\lambda_3 = \frac{|\mathbf{g}_2 \cdot \mathbf{h}_2|}{|\mathbf{A} \cdot \mathbf{h}_2|^2} \,. \tag{2.77}$$

A system of $N$ equations requires a total of $N$ such steps to determine the solution vector $\mathbf{x}$.

EXAMPLE: As already mentioned, the CG method is most appropriate for large systems of equation with a sparsely inhabited matrix $\mathbf{A}$. But the necessary manipulations may be demonstrated using the 2-dimensional example we have used before. Let once more

$$\mathbf{A} = \begin{pmatrix} 3 & 1 \\ 2 & 4 \end{pmatrix} ; \ \mathbf{b} = \begin{pmatrix} 3 \\ 2 \end{pmatrix} ; \ \text{and} \ \mathbf{x}_0 = \begin{pmatrix} 1.2 \\ 0.2 \end{pmatrix}$$

The gradient vector at $\mathbf{x}_0$ is

$$\mathbf{g}_0 = -\mathbf{A}^T \cdot [\mathbf{A} \cdot \mathbf{x}_0 - \mathbf{b}] = - \begin{pmatrix} 4.8 \\ 5.6 \end{pmatrix}$$

and

$$\lambda_1 = \frac{|\mathbf{g}_0|^2}{|\mathbf{A} \cdot \mathbf{g}_0|^2} = 0.038$$

so that

$$\mathbf{x}_1 = \mathbf{x}_0 + \lambda_1 \mathbf{g}_0 = \begin{pmatrix} 1.017 \\ -0.014 \end{pmatrix}$$

**Conjugate gradient technique:**

1. Let $P_0$ (with the position vector $\mathbf{x}_0$) be the starting point of the search; the local gradient at $P_0$ is

$$\mathbf{g}_0 \equiv -\nabla f(\mathbf{x}_0) = -\mathbf{A}^T \cdot [\mathbf{A} \cdot \mathbf{x}_0 - \mathbf{b}] \qquad (2.68)$$

The next "low point" $P_1$ is then situated at

$$\mathbf{x}_1 = \mathbf{x}_0 + \lambda_1 \, \mathbf{g}_0 \qquad (2.69)$$

with

$$\lambda_1 = \frac{|\mathbf{g}_0|^2}{|\mathbf{A} \cdot \mathbf{g}_0|^2} \,. \qquad (2.70)$$

2. From $P_1$ we proceed *not* along the local gradient

$$\mathbf{g}_1 = -\mathbf{A}^T \cdot [\mathbf{A} \cdot \mathbf{x}_1 - \mathbf{b}] \qquad (2.71)$$

but along the gradient conjugate to $\mathbf{g}_0$, i.e.

$$\mathbf{h}_1 = \mathbf{g}_1 - \frac{\mathbf{g}_1 \cdot \mathbf{A} \cdot \mathbf{g}_0}{\mathbf{g}_0 \cdot \mathbf{A} \cdot \mathbf{g}_0} \, \mathbf{g}_0 \,. \qquad (2.72)$$

The low point along this path is at

$$\mathbf{x}_2 = \mathbf{x}_1 + \lambda_2 \, \mathbf{h}_1 \,, \qquad (2.73)$$

with

$$\lambda_2 = \frac{|\mathbf{g}_1 \cdot \mathbf{h}_1|}{|\mathbf{A} \cdot \mathbf{h}_1|^2} \qquad (2.74)$$

Figure 2.2: The CG method

Similarly we find from 2.71-2.73

$$\mathbf{g}_1 = \begin{pmatrix} -0.063 \\ 0.054 \end{pmatrix}, \quad \mathbf{h}_1 = \begin{pmatrix} -0.071 \\ 0.044 \end{pmatrix}, \quad \lambda_2 = 0.231,$$

and thus

$$\mathbf{x}_2 = \begin{pmatrix} 1.000 \\ -0.004 \end{pmatrix}.$$

It took us just two steps to find the solution to the 2-dimensional system $\mathbf{A} \cdot \mathbf{x} = \mathbf{b}$. If $\mathbf{A}$ were a $N \times N$ matrix, $N$ such steps would be necessary.


## 2.3     Eigenvalues and Eigenvectors

Given a matrix $\mathbf{A}$, the physicist who needs the eigenvalues $\lambda_i$ defined by

$$|\mathbf{A} - \lambda_i \, \mathbf{I}| = 0 \quad (i = 1, \ldots N) \tag{2.78}$$

and the corresponding eigenvectors $\mathbf{a}_i$,

$$[\mathbf{A} - \lambda_i \, \mathbf{I}] \cdot \mathbf{a}_i = 0, \tag{2.79}$$

will normally make use of one of the various standard subroutine packages. In the NAG library, for instance, these would be routines with names like F01xxx, F02xxx; the respective ESSL routine would be SGEEV.

In some situations, however, it is sufficient to determine only a few – typically the largest – eigenvalues and the associated eigenvectors. Examples are Courant and Hilbert's stability analysis of numerical algorithms for the solution of differential equations (Section 4.1, [GEAR 71]) and quantum mechanical perturbation theory ([KOONIN 85, MCKEOWN 87].) In such cases it is obviously not a good idea to use the too comprehensive standard routines. Rather one will apply one of the following iterative procedures.


### 2.3.1     Largest Eigenvalue and Related Eigenvector

The $N$ eigenvectors $\mathbf{a}_i$ of a matrix $\mathbf{A}$ may be viewed as the base vectors of a coordinate system. An arbitrary $N$-vector $\mathbf{x}_0$ is then represented by

$$\mathbf{x}_0 = \sum_{i=1}^{N} c_i \, \mathbf{a}_i \tag{2.80}$$

with suitable coefficients $c_i$. Let us assume that $\mathbf{x}_0$ contains a non-vanishing component $c_m$ along that eigenvector $\mathbf{a}_m$ which corresponds to the largest (by absolute value) eigenvalue $\lambda_m$. Now multiply $\mathbf{x}_0$ several times by $\mathbf{A}$, each time normalizing the result:

$$\mathbf{x}_k{}' = \mathbf{A} \cdot \mathbf{x}_{k-1} \quad \Longrightarrow \quad \mathbf{x}_k = \frac{\mathbf{x}_k{}'}{|\mathbf{x}_k{}'|} \tag{2.81}$$

After a few iterations we have

$$\mathbf{x}_k \propto \sum_{i=1}^{N} c_i \lambda_i^k \, \mathbf{a}_i \approx c_m \lambda_m^k \, \mathbf{a}_m \tag{2.82}$$

This is to say that the iterated vector will be dominated by the $\mathbf{a}_m$-component. The result is therefore a unit vector with direction $\mathbf{a}_m$. The eigenvalue $\lambda_m$ may be obtained from either of the following formulae:

$$\lambda_m = \frac{x'^{(k)}_\beta}{x^{(k-1)}_\beta} \quad \text{or} \quad \lambda_m = \mathbf{x}_{k+1} \cdot \mathbf{x}'_k \tag{2.83}$$

where $x'^{(k)}_\beta$ denotes any cartesian component of the – still unnormalized – vector $\mathbf{x}_k{}'$. Of course, the equations 2.83 apply only when all components except $\mathbf{a}_m$ have become negligible.

EXAMPLE: Once more, let

$$\mathbf{A} = \begin{pmatrix} 3 & 1 \\ 2 & 4 \end{pmatrix}$$

and choose as the starting vector

$$\mathbf{x}_0 = \begin{pmatrix} \sqrt{2}/2 \\ \sqrt{2}/2 \end{pmatrix}$$

The iterated and normalized vectors (see equs. 2.81) are

$$\mathbf{x}_1 = \begin{pmatrix} 0.555 \\ 0.832 \end{pmatrix} ; \ \mathbf{x}_2 = \begin{pmatrix} 0.490 \\ 0.872 \end{pmatrix} ; \ \mathbf{x}_3 = \begin{pmatrix} 0.464 \\ 0.886 \end{pmatrix} ; \dots$$

From $\mathbf{x}_3$ and the still unnormalized

$$\mathbf{x}_4{}' = \begin{pmatrix} 2.279 \\ 4.471 \end{pmatrix}$$

we find, using the first of equs. (2.83), $\lambda_m = 4.907$. The exact solution of the problem is

$$\mathbf{a}_m = \begin{pmatrix} 0.45 \\ 0.90 \end{pmatrix} \quad \text{and} \quad \lambda_m = 5 \,.$$

## 2.3.2 Arbitrary Eigenvalue/-vector: Inverse Iteration

The foregoing recipe may be modified so as to produce that eigenvalue $\lambda_n$ which is nearest to some given number $\lambda$. Again we set out from an arbitrary vector $x_0$. The iterative procedure is now defined by

$$\mathbf{x}_k{}' = [\mathbf{A} - \lambda\,\mathbf{I}]^{-1} \cdot \mathbf{x}_{k-1} \qquad \Longrightarrow \qquad \mathbf{x}_k = \frac{\mathbf{x}_k{}'}{|\mathbf{x}_k{}'|} \tag{2.84}$$

It is easy to see that after a few iterations the vector

$$\mathbf{x}_k \propto \sum_{i=1}^{N} c_i\,[\lambda_i - \lambda]^{-k}\,\mathbf{a}_i \tag{2.85}$$

contains almost exclusively the component corresponding to $\lambda_n$:

$$\mathbf{x}_k \to c_n\,[\lambda_n - \lambda]^{-k}\,\mathbf{a}_n \tag{2.86}$$

$\lambda_n$ itself may then be evaluated using either one of the obvious relations

$$\lambda_n - \lambda = \frac{x_\beta{}^{(k)}}{x'_\beta{}^{(k-1)}} \qquad \text{or} \qquad \lambda_n = \lambda + \frac{1}{\mathbf{x}_{k-1} \cdot \mathbf{x}'_k} \tag{2.87}$$

EXAMPLE: With the same sample matrix as before and an estimated value $\lambda = 1$ the iteration matrix in 2.84 is given by

$$[\mathbf{A} - \lambda\,\mathbf{I}]^{-1} = \frac{1}{4}\begin{pmatrix} 3 & -1 \\ -2 & 2 \end{pmatrix}$$

Starting out from

$$\mathbf{x}_0 = \begin{pmatrix} 1 \\ 0 \end{pmatrix}$$

we find for the iterated, normalized vectors

$$\mathbf{x}_1 = \begin{pmatrix} 0.832 \\ -0.555 \end{pmatrix} ; \mathbf{x}_2 = \begin{pmatrix} 0.740 \\ -0.673 \end{pmatrix} ; \mathbf{x}_3 = \begin{pmatrix} 0.715 \\ -0.699 \end{pmatrix} ; \mathbf{x}_4 = \begin{pmatrix} 0.709 \\ -0.705 \end{pmatrix} \cdots$$

The next vector is, before normalization,

$$\mathbf{x}_5{}' = \begin{pmatrix} 0.708 \\ -0.707 \end{pmatrix}$$

so that $\mathbf{x_5}' \cdot \mathbf{x_4} = 1.0015$. Using the first of equs. 2.87 we have $\lambda_n = 2.001$. The exact eigenvalues of $\mathbf{A}$ are 5 and 2; the eigenvector corresponding to $\lambda = 2$ is

$$ \mathbf{a} = \left( \begin{array}{c} 0.707 \\ -0.707 \end{array} \right) $$

In going through the above exercise we are reminded that – see equ. 2.84 – a matrix inversion is required. This is in contrast to the direct iteration 2.81. Inverse iteration is therefore appropriate only if no more than a few eigenvalues/-vectors of a large matrix are needed. In other cases it may be advisable after all to invoke the well-optimized standard routines.

## 2.4    Sample Applications

Within physics the most prominent areas of application of linear algebra are continuum theory and quantum mechanics. In the theory of continua, systems of linear equations occur whenever one of the partial differential equations that abound there is discretized. This will be discussed at length in Chapter 5 (Partial Differential Equations) and 8 (Hydrodynamics); here we present just two examples (Secs. 2.4.1 and 2.4.2). In quantum mechanics, linear systems are equally ubiquitous. We will provide an example (Sec. 2.4.3) and for further information refer the reader to the truly extensive literature which in this instance is to be found mostly in the neighboring realm of quantum chemistry.

Further applications of linear algebra will be treated in Chapter 3 (Stochastics).

### 2.4.1    Diffusion and Thermal Conduction

In Section 1.4 we have shown how to discretize the diffusion equation (or equation of thermal conduction) by applying the DNGF and DDST formulae. Without giving arguments we simply used the DDST approximation *at time $t_n$*, writing

$$ \frac{\partial u(x,t)}{\partial x^2} \approx \frac{\delta_i^2 u_i^n}{(\Delta x)^2} \tag{2.88} $$

In this manner we arrived at the "FTCS-"formula. However, with no less justification we may use the same spatial differencing *at time $t_{n+1}$*,

$$ \frac{\partial u(x,t)}{\partial x^2} \approx \frac{\delta_i^2 u_i^{n+1}}{(\Delta x)^2} \tag{2.89} $$

This leads us to the "implicit scheme of first order"

$$ \frac{1}{\Delta t}[u_i^{n+1} - u_i^n] = \frac{D}{(\Delta x)^2}[u_{i+1}^{n+1} - 2u_i^{n+1} + u_{i-1}^{n+1}] \tag{2.90} $$

which may be written, using $a \equiv D\,\Delta t/(\Delta x)^2$,

$$-au_{i-1}^{n+1} + (1+2a)u_i^{n+1} - au_{i+1}^{n+1} = u_i^n \tag{2.91}$$

for $i = 1, \ldots N-1$. Once more fixing the boundary values $u_0$ and $u_N$ we may write this system of equations in matrix form, thus:

$$\mathbf{A} \cdot \mathbf{u}^{n+1} = \mathbf{u}^n \tag{2.92}$$

where

$$\mathbf{A} \equiv \begin{pmatrix} 1 & 0 & 0 & . & . & 0 \\ -a & 1+2a & -a & 0 & . & 0 \\ 0 & . & & . & . & 0 & . \\ . & . & & . & . & . \\ . & . & & . & 0 & 0 & 1 \end{pmatrix} \tag{2.93}$$

It is now an easy matter to invert this tridiagonal system by the recursion scheme of Sec. 2.1.4.

EXERCISE: Solve the problem of Sec. 1.4 (one-dimensional thermal conduction) by applying the implicit scheme in place of the FTCS method. Use various values of $\Delta t$ (and therefore $a$.) Compare the efficiencies and stabilities of the two methods.

## 2.4.2   Potential Equation

In a later section we will concern ourselves in loving detail with partial differential equations of the form

$$\frac{\partial^2 u}{\partial x^2} + \frac{\partial^2 u}{\partial y^2} = -\rho \tag{2.94}$$

According to general typology we are here dealing with an *elliptic* PDE. The electrostatic potential produced by a charge density $\rho(x,y)$ obeys this equation, which was first formulated by Poisson. The equation can be solved uniquely only if the values of the solution $u(x,y)$ are given along a boundary curve $C(x,y) = 0$ (Dirichlet boundary conditions,) or if the derivatives $(\partial u/\partial x, \partial u/\partial y)$ are known along such a curve (Neumann boundary conditions.)

By introducing finite differences $\Delta x = \Delta y$ we derive from 2.94 the difference equations

$$\frac{1}{(\Delta x)^2} \left[ u_{i+1,j} - 2u_{i,j} + u_{i-1,j} + u_{i,j+1} - 2u_{i,j} + u_{i,j-1} \right] = -\rho_{i,j} \tag{2.95}$$

$$i = 1, \ldots N;\, j = 1, \ldots M$$

Combining the $N$ row vectors $\{u_{i,j};\ j = 1, \ldots M\}$ sequentially to a vector $\mathbf{v}$ of length $N.M$ we may write these equations in the form

$$\mathbf{A} \cdot \mathbf{v} = \mathbf{b} \tag{2.96}$$

where $\mathbf{A}$ is a sparse matrix, and where the vector $\mathbf{b}$ contains the charge density $\rho$ and the given boundary values of the potential function $u$ (see Section 5.3).

Any of the methods of solution which we have discussed in this chapter may now be applied to equ. 2.96. Actually the relaxation methods and the ADI technique are the most popular procedures. In addition there are specialized methods that are tailored to the potential equation (see Secs. 5.3.3 and 5.3.4).

## 2.4.3  Electronic Orbitals

The wave function of the electrons in a molecular shell is frequently expressed as a linear combination of atomic orbitals (MO-LCAO approximation):

$$\Psi = \sum_i a_i \psi_i \tag{2.97}$$

where $\psi_i$ is the wave function of the shells contributing to the molecular bond. Applying the Schroedinger equation to this linear combination one finds

$$\sum_i a_i H \psi_i = E \sum_i a_i \psi_i \tag{2.98}$$

and further

$$\sum_i a_i H_{ji} = E \sum_i a_i S_{ji} \tag{2.99}$$

with

$$H_{ji} \equiv \langle j|H|i\rangle = \int \psi_j^* H \psi_i d\mathbf{r} \; ; \; S_{ji} = \langle j|i\rangle = \int \psi_j^* \psi_i d\mathbf{r} \tag{2.100}$$

Equ. 2.99 is just a generalized eigenvalue problem of the form

$$\mathbf{H} \cdot \mathbf{a} = E\mathbf{S} \cdot \mathbf{a} \tag{2.101}$$

which may be solved using the procedures described above.

A particularly transparent example for the application of the LCAO method is the Hueckel theory of planar molecules; see, e.g., [MCKEOWN 87].

# Chapter 3

# Stochastics

*John von Neumann would later toss the dice using computers*

The idea to include chance in a model of reality may be traced back even to antiquity. The Epicuraeans held that the irregular motion of atoms arises because individual atoms stray "without cause" from their straight paths. Such views necessarily elicited angry opposition from those scholars who believed in predetermination. And even the "philosophy professor" Cicero, himself an eminent critic of the exaggerated causality doctrine of the Stoics, comments caustically:

> "So what new cause is there in nature to make the atoms swerve? Or do they draw lots among themselves which will swerve and which not? Or do they swerve by a minimum interval and not by a larger one, or why do they swerve by one minimum and not by two or three? This is wishful thinking, not argument." [CICERO -44]

In fact, the same argument is still going on today – albeit with a slightly different vocabulary. Just remember the dispute between the mechanists and the champions of free will, the passionate discussion around Jaques Monod's book "Chance and Necessity", or the laborious struggle of philosophy with quantum mechanical uncertainty.

With becoming epistemological humility we will refrain from trying to explain the whole world at once. Let us content ourselves with modelling a small subsection of physical reality. But then the boundary of our subsystem will be permeable to influences – fields, forces, collisions etc. – originating in the encompassing system. To avoid having to include the larger system in the description we will replace its influence on the subsystem by suitably chosen "accidental" fields, forces, collisions etc. Just this is the basis of stochastic methods in physics.

Let us reflect for a moment on the interrelated concepts "statistical" and "stochastic." A *statisticus* was the administrator of a Roman country estate or a manufacture. It was his task to extract the regularities – like the total amount of wheat brought in – hidden in the everyday turmoil. This is just what a statistician does: out of a heap of more or less irregular data he distills the essential parameters – mean, standard deviation and such.

In contrast, *stochastic* means simply *irregular* or *arbitrary*. While in statistics we aim to extract the regular from the irregular, in stochastics we put the irregular to work – for instance in "trying out" many possible states of a model system.

There are various ways by which to account for the irregular influence of the environment upon the modelled subsystem. In Boltzmann's kinetic theory of gases and in Smoluchowski's description of diffusional motion random forces do not appear explicitly. Rather, they are accounted for *modo statistico* by way of certain mathematical assumptions on the probability density in phase space – molecular chaos, detailed balance etc.

Alternatively, the diffusive motion of a particle may be described in terms of a stochastic equation of motion in which the factor of chance is represented explicitly in the form of a *stochastic force*. In 1907 Paul Langevin postulated the following equation for the motion of a Brownian particle:

$$m \frac{d^2}{dt^2} \mathbf{r}(t) = -\gamma \mathbf{v}(t) + \mathbf{S}(t) \,, \tag{3.1}$$

Here $-\gamma \mathbf{v}$ is the decelerating viscous force acting on the particle as it moves through the surrounding fluid, and $\mathbf{S}$ is the stochastic force which arises from the irregular impacts of the fluid's molecules. Incidentally, it took kinetic theorists more than sixty years to come up with a strict derivation of Langevin's equation [MAZUR 70].

To produce a solution to this equation of motion we must first of all draw the actual value of the random force $\mathbf{S}$ by some "gambling" procedure (O Cicero!). The mean value of each Cartesian component of $\mathbf{S}$ must of course be zero, and the variance is closely related to the viscosity $\gamma$ and the temperature of the fluid. In Chapter 6 we will take a closer look at this method of "Stochastic dynamics." For the moment let us note that the crucial step in this technique is the sampling of certain random variates. In fact, we may take it as an operational definition of stochastic methods in computational physics that in applying such methods one has to call a random number generator.

In Gibbs' version of statistical mechanics one studies, in place of one single model system, a large number of inaccurate copies of that system. Each member of the so defined "ensemble" differs in detail from the others, with the variance of these deviations being known. Once more, chance appears only in an implicit manner, namely in the form of *statistical* assumptions. Nevertheless there exists a decidedly *stochastic* method for evaluating averages over an ensemble: the Monte Carlo method. Here the ensemble is constructed step by step, by producing a sequence of "erroneous copies" of a given model system. At each copying step the manner and extent of deviation from the preceding copy is sampled; this is called a *random walk* through phase space. In a later chapter (6) the statistical-mechanical Monte Carlo method will be explained in more detail. But we note here that what is obviously needed once more is a "loaded die" – that is, a random number generator that produces a sequence of numbers with certain desired statistical properties.

Depending on the specific kind of application we will need random variates with different probability distributions. The most simple task is the production of equidistributed random numbers. But the access to all other distributions is passing through the equidistribution as well. The following section is therefore devoted to the methods that enable us to construct sequences of equidistributed random variates. To proceed to other distributions one may then use the *transformation method* (see Section 3.2.2), invoke the *rejection method* (Section 3.2.4), or set out on a *random walk* (Section 3.3.5).

# 3.1    Equidistributed Random Variates

The correct name, of course, is "pseudorandom" numbers, since any numerical algorithm for producing a sequence of numbers is necessarily deterministic. However, we will be quite satisfied if the numbers thus produced, when submitted to certain statistical tests, are free of undesirable regularities [MARSAGLIA 90]. In that case we will overlook the fact that as a rule they do not come from a "truly random" process.[1] One requirement, however, must hold: the relevant algorithms should be very fast, since in the course of a Monte Carlo calculation or a diffusional random walk we need large amounts of random numbers.

## 3.1.1    Linear Congruential Generators

The classic method for producing a sequence of homogeneously distributed random numbers is defined by the recursive prescription

$$I_{n+1} = [a I_n + b] \bmod m \qquad\qquad (3.2)$$

---

[1]In fact, there are increasingly successful attempts to construct "physical" random number generators which may be based on thermal noise in resistors or on quantum phenomena [STAUFFER 89, JENNEWEIN 00].

(see [ABRAMOWITZ 65], [PRESS 86], [KNUTH 69]). Here, $a$ is some (odd) multiplicative factor, $m$ is the largest integer that may be represented by the particular computer (usually $m = 2^{32}$ or such), and $b$ is relatively prime with respect to $m$ (i.e. $b$ and $m$ have no common factor).

The numbers produced in this manner are homogeneously distributed over the whole range of representable integers. A sequence of random numbers $x_n$ of type *real*, equidistributed over the interval $(0,1)$, may be obtained by dividing $I_n$ by $m$.

Most high-level programming languages contain some internal routine based on this technique. These routines are usually called by names like RAND, RND, RAN etc. (The word *random*, incidentally, stems from ancient French, where *randon* meant impulsiveness or impetuosity.) The first number in the sequence, the – odd-numbered – "seed" $I_0$, may often be chosen by the user.

Statistical scrutiny shows that the random numbers produced in this way are not very "good." While a histogram of their relative frequencies looks quite inconspicuous, there are undesirable serial correlations of the type

$$\langle x_n \, x_{n+k} \rangle \neq 0 \; ; \quad k = 1, 2, \ldots \tag{3.3}$$

It depends on the particular application whether such autocorrelations are acceptable or not. For instance, every 3 successive $x_n$ might be used as cartesian coordinates of a point inside the unit cube. In that case one would find that the points would be confined to a discrete manifold of parallel planes ([COLDWELL 74]).

There is a simple and economical trick to cleanse the internal random number generator from its serial correlations. The procedure to follow is described in Figure 3.1 [PRESS 86].

## 3.1.2  Shift Register Generators

There are several names for this group of techniques. One may encounter them as "Tausworthe" or "XOR" generators, or as the method of "primitive polynomials." Originally these methods were designed for the production of *random bits*, but one may always generate 16, 32, etc. bits at a time and combine them to a computer word.

The procedure is very simple. Assuming that $n$ random bits $b_1, b_2, ..b_n$ are already given, we apply a recursive rule of the form

$$b_{n+1} = b_k \oplus b_m \oplus \ldots \oplus b_n \, , \tag{3.4}$$

to find another random bit. Here, $k < m < .. < n$, and $\oplus$ denotes the logical operation "exclusive or" (XOR) which yields the result 1 only if any one, but not both, of the two operands equals 1.

The properties of the generator 3.4 will obviously depend on the actual combination of indices $(k, m, \ldots, n)$. This is not the place to reproduce the analysis

---

**"Erasing tracks:"**

1. Produce a list $RLIST(i)$ of $Z$ equidistributed random numbers $x_i \in (0, 1)$; $i = 1 \ldots Z$. $Z$ should be prime and no less than about 100, e.g. $Z = 97$.

2. Sample an additional random number $y$ in $(0, 1)$.

3. Determine a pointer index $j \in [1, Z]$ according to

$$j = 1 + \text{int}(y \cdot Z)$$

   $(\text{int}(r) \ldots$ largest integer smaller than the real number $r$.)

4. Use the element $RLIST(j)$ corresponding to $j$ as the output random number.

5. Put $y = RLIST(j)$ and replace $RLIST(j)$ by a new random number $\in (0, 1)$; return to (3).

---

Figure 3.1: Removal of autocorrelations in simple congruential generators

leading to a class of optimal index combinations. Suffice it to refer to the theory of "primitive polynomials modulo 2" [TAUSWORTHE 65, NIEDERREITER 82]. These are a subset of all polynomials whose coefficients and variables may take on the values 0 or 1 only:

$$P(x; k, m, \ldots n) = 1 + x^k + x^m + \ldots + x^n ; \; x = 0 \text{ or } 1 \qquad (3.5)$$

A table of primitive polynomials modulo 2 may be found in [PRESS 86], p. 212.

We may use any such polynomial modulo 2, be it primitive or not, to define a recursion prescription of the form (3.4). The specific advantage of *primitive* polynomials is that the recursion procedures defined by them exhibit a certain kind of "exhaustive" property. Starting such a recursion with an arbitrary combination of $n$ bits (except $0 \ldots 0$), *all possible* configurations of $n$ bits will be realized just once before a new cycle begins.

EXAMPLE: The sequence (1,3) defines a primitive polynomial modulo 2. Starting with the arbitrary bit combination 101 we obtain by applying the prescription

$$b_4 \;\; = \;\; b_3 \oplus b_1$$

$$\cdots$$

$$b_s \;\; = \;\; b_{s-1} \oplus b_{s-3} ; \; s = 4, 5, ..$$

the sequence, reading from left to right,

$$101\,001\,110\,100\,111\,010\,011\,101 \, \cdots$$

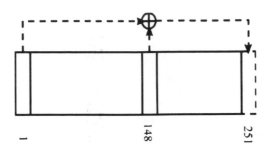

Figure 3.2: Kirkpatrick-Stoll prescription

It is evident that indeed all possible 3-bit groups (except 000) occur before the sequence repeats.

Primitive trinomials of the form

$$P(x; m, n) = 1 + x^m + x^n \tag{3.6}$$

yield recursion formulae which require only one XOR operation per step:

$$b_s = b_{s-m} \oplus b_{s-n}; \ s = n + 1, \ldots \tag{3.7}$$

A specific prescription of this type which has been developed and tested by Kirkpatrick and Stoll [KIRKPATRICK 81, KALOS 86] makes use of the indices $m = 103$ and $n = 250$.

In all high-level programming languages the XOR command may be applied to arguments of the type *integer* as well. The code line

$$I_s = I_{s-103} \oplus I_{s-250} \tag{3.8}$$

which corresponds to the "R250" algorithm of Kirkpatrick-Stoll, means that the two integers on the right-hand side are to be submitted bit by bit to the XOR operation. Again, random numbers of type *real* within the range $(0, 1)$ may be obtained by machine-specific normalization.

To start a generator of this type one must first produce 250 random integers. For this purpose a linear congruential generator may be used. To keep the storage requirements within bounds while applying a recursion like 3.8 one will provide for some sort of cyclic replacement of register contents.

Overviews on modern random number generators, in particular on Tausworthe algorithms and the related Fibonacci generators, are given in [JAMES 90] and [MARSAGLIA 90]. More recent developments are reviewed in [GUTBROD 99].

## 3.2    Other Distributions

### 3.2.1    Fundamentals

Before describing the methods for producing random numbers with arbitrary sta-tistical distributions we have to clarify a few basic concepts:

**Distribution function:** Let $x$ be a real random variate with a range of values $(a, b)$. By *distribution function* we denote the probability that $x$ be less than some given value $x_0$:

$$P(x_0) \equiv \mathcal{P}\{x < x_0\} \tag{3.9}$$

A common example in which $a = -\infty$ and $b = \infty$ is the Gaussian, or normal, distribution

$$P(x_0) = \frac{1}{\sqrt{2\pi}} \int_{-\infty}^{x_0} dx\, e^{-x^2/2} \tag{3.10}$$

The function $P(x)$ is monotonous and non-decreasing, with $P(a) = 0$ and $P(b) = 1$. The distribution function is dimensionless: $[P(x)] = 1$.

**Probability density:** The *probability (or distribution) density* $p(x)$ is defined by the identity

$$p(x_0)\, dx \equiv \mathcal{P}\{x \in [x_0, x_0 + dx]\} \equiv dP(x_0) \tag{3.11}$$

Thus $p(x)$ is simply the differential quotient of the distribution function:

$$p(x) = \frac{dP(x)}{dx}, \quad \text{i.e. } P(x_0) = \int_a^{x_0} p(x)\, dx \tag{3.12}$$

The dimension of $p(x)$ equals the inverse of the dimension of $x$:

$$[p(x)] = \frac{1}{[x]} \tag{3.13}$$

In the above example $p(x)$ would be

$$p(x) = \frac{1}{\sqrt{2\pi}} e^{-x^2/2} \tag{3.14}$$

If $x$ may take on discrete values $x_\alpha$ only, with $\Delta x_\alpha \equiv x_{\alpha+1} - x_\alpha$, we use the notation

$$p_\alpha \equiv p(x_\alpha)\, \Delta x_\alpha \tag{3.15}$$

for the probability of the event $x = x_\alpha$. This quantity $p_\alpha$ is by definition dimensionless, in spite of its being related to the probability density $p(x)$ of a continuous random variate. The discrete variant of the (cumulative) distribution *function* is simply

$$P_\beta = \sum_{\alpha=1}^{\beta} p_\alpha \tag{3.16}$$

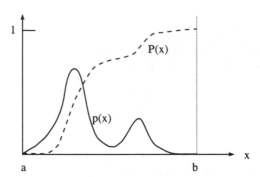

Figure 3.3: Distribution function and density

**Statistical (in)dependence:** Two random variates $x_1, x_2$ are said to be *statistically independent* or *uncorrelated* if the density of the *compound probability* – that is, the probability for $x_1$ and $x_2$ occuring simultaneously – equals the product of the individual probabilities:

$$p(x_1, x_2) = p(x_1)\, p(x_2) \tag{3.17}$$

In practical applications this means that one may sample each of the two variates from its own distribution, regardless of the actual value of the other variable.

By *conditional probability density* we denote the quantity

$$p(x_2|x_1) \equiv \frac{p(x_1, x_2)}{p(x_1)} \tag{3.18}$$

(For uncorrelated $x_1, x_2$ we have $p(x_2|x_1) = p(x_2)$).

The density of the *marginal distribution* gives the density of one of the two variables, irrespective of the actual value of the other one; in other words, it is an integral over the range of values of that other variate:

$$p(x_2) \equiv \int_{a_1}^{b_1} p(x_1, x_2)\, dx_1 \tag{3.19}$$

**Moments of a probability density:** These are the quantities

$$\langle x^n \rangle \equiv \int_{a}^{b} x^n p(x)\, dx \tag{3.20}$$

In the case of two (or more) random variates the definition is to be suitably generalized, as in

$$\langle x_1^m x_2^n \rangle \equiv \int_{a_1}^{b_1} \int_{a_2}^{b_2} x_1^m\, x_2^n\, p(x_1, x_2)\, dx_1\, dx_2 \tag{3.21}$$

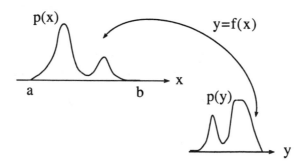

Figure 3.4: Transformation of the probability density

In particular the quantity $\langle x_1 x_2 \rangle$ is called the *cross correlation* or *covariance* of $x_1$ and $x_2$. If the two variates are statistically independent (uncorrelated), we have $\langle x_1 x_2 \rangle = \langle x_1 \rangle \langle x_2 \rangle$.

**Transformation of probability densities:** From equ. 3.11 we may easily derive a prescription for the transformation of a density $p(x)$ upon substitution of variables $x \leftrightarrow y$. Given a bijective mapping $y = f(x)$; $x = f^{-1}(y)$, and given the density $p(x)$, the conservation of probability requires

$$|dP(y)| = |dP(x)| \tag{3.22}$$

(The absolute value occurs here since we have not required the function $f(x)$ to be increasing.) It follows that

$$|p(y)\,dy| = |p(x)\,dx| \tag{3.23}$$

or

$$p(y) = p(x)\left|\frac{dx}{dy}\right| = p[f^{-1}(y)]\left|\frac{df^{-1}(y)}{dy}\right| \tag{3.24}$$

Incidentally, the relation 3.24 holds for any kind of density, such as mass or spectral densities, not only for probability densities.

EXAMPLE: The spectral density of black body radiation is usually written in terms of the angular frequency $\omega$:

$$I(\omega) = \frac{\hbar\omega^3}{\pi c^3}\frac{1}{e^{\hbar\omega/kT} - 1} \tag{3.25}$$

If we prefer to give the spectral density in terms of the wave length $\lambda \equiv 2\pi c/\omega$, we have from 3.24

$$I(\lambda) = I[\omega(\lambda)]\left|\frac{d\omega}{d\lambda}\right| = \frac{\hbar}{\pi c^3}\left(\frac{2\pi c}{\lambda}\right)^3\frac{1}{e^{(hc/\lambda)/kT} - 1}\left(\frac{2\pi c}{\lambda^2}\right) \tag{3.26}$$

---

**Transformation method:**

Let $p(x)$ be a desired density, with a corresponding distribution function $y = P(x)$. The inverse of the latter, $P^{-1}(y)$, is assumed to be known.

- Sample $y$ from an equidistribution in the interval $(0, 1)$.

- Compute $x = P^{-1}(y)$.

The variable $x$ then has the desired probability density $p(x)$.

---

Figure 3.5: Transformation method

EXERCISE: A powder of approximately spherical metallic grains is used for sintering. The diameters of the grains obey a normal distribution with $\langle d \rangle = 2\mu m$ and $\sigma = 0.25\mu m$. Determine the distribution of the grain volumes.

## 3.2.2    Transformation Method

Let us now return to our task of generating random numbers $x$ with some given probability density (or relative frequency) $p(x)$. We will first try to find a bijective mapping $y = f(x)$ such that the distribution of $y$ is homogeneous, i.e. $p(y) = c$. By the transformation law for densities (read backwards) we will then have

$$p(x) = c \left| \frac{dy}{dx} \right| = c \left| \frac{df(x)}{dx} \right| \tag{3.27}$$

This means that in order to serve our purpose the mapping $y = f(x)$ should obey

$$\left| \frac{df(x)}{dx} \right| = \frac{1}{c} p(x) \tag{3.28}$$

It is easy to see that the mapping

$$f(x) \equiv P(x) \tag{3.29}$$

fulfills this condition, and that $c = 1$. This solves our problem: all we have to do now is sample $y$ from an equidistribution $\in [0, 1]$ and compute the inverse $x = P^{-1}(y)$ (see Figs. 3.5, 3.6).

EXAMPLE: Let

$$p(x) = \frac{1}{\pi} \frac{1}{1 + x^2} \quad \text{(Lorentzian)} \tag{3.30}$$

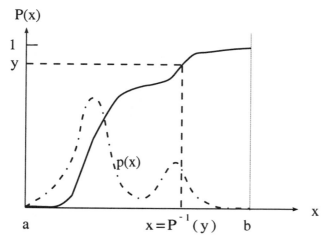

Figure 3.6: Transformation method: geometrical interpretation

be the desired density in the interval ($\pm\infty$). The integral function of $p(x)$ is then $y = P(x) = 1/2 + (1/\pi)$ arctan $x$, and the inverse of that is $P^{-1}(y) = \tan[\pi(y - 1/2)]$. The prescription for producing random variates $x$ distributed according to 3.30 is therefore

- Sample $y$ equidistributed in $(0, 1)$.

- Compute $x = \tan[\pi(y - \frac{1}{2})]$.

A geometrical interpretation of this procedure may be found from Fig. 3.6. If $y$ is sampled from a homogeneous distribution $\in (0, 1)$ and transformed into an $x$-value using $x = P^{-1}(y)$, then those regions of $x$ in which $P(x)$ is steeper are obviously hit more frequently. The slope of $P(x)$, however, is just equal to $p(x)$, so that $x$-values with large $p(x)$ are indeed sampled more often than others.

Sometimes the primitive function $P(x)$ of the given density $p(x)$ is not an analytical function, or if it is, it may not be analytically invertible. In such cases one may take recourse to approximation and interpolation formulae, or else use the "rejection method" to be described later on.

### 3.2.3    Generalized Transformation Method:

The foregoing considerations on the transformation of distribution densities are valid not only for a single random variate $x$, but also for vectors $\mathbf{x} = (x_1, \ldots, x_n)$ made up of several variables. Let $\mathbf{x}$ be such a vector defined within an $n$-dimensional region $D_x$, and let $\mathbf{y} = \mathbf{f}(\mathbf{x})$ be a bijective mapping onto a corresponding region $D_y$ (see Fig. 3.7). Again invoking conservation of probability we find

$$p(\mathbf{y}) = p(\mathbf{x}) \left| \frac{\partial \mathbf{x}}{\partial \mathbf{y}} \right| , \tag{3.31}$$

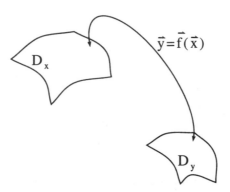

Figure 3.7: Transformation in higher dimensions

where $|\partial \mathbf{x}/\partial \mathbf{y}|$ is now the Jacobi determinant of the transformation $\mathbf{x} = \mathbf{f}^{-1}(\mathbf{y})$.

## Normal Distribution (Box-Muller Method)

An important application of the generalized transformation method is the follow-ing, widely used technique for generating normal random variates.[2] Let

$$p(\mathbf{x}) \equiv p(x_1, x_2) = \frac{1}{2\pi} e^{-(x_1^2 + x_2^2)/2}, \tag{3.32}$$

be the common density of two uncorrelated normal variates. By introducing polar coordinates $(r, \phi)$ instead of $(x_1, x_2)$ we find

$$p(r, \phi) = \left[r e^{-r^2/2}\right] \left[\frac{1}{2\pi}\right] \equiv p(r)\, p(\phi) \tag{3.33}$$

Thus the variable $y_2 \equiv \phi/2\pi$ is already homogeneously distributed in $(0,1)$ and statistically independent of $r$, and we are left with the problem of reproducing the density $p(r)$. The quantity

$$y_1 \equiv P(r) = \int_0^r p(r')\, dr' = 1 - e^{-r^2/2} \tag{3.34}$$

is equidistributed in $(0,1)$. Consequently, $1 - y_1$ is equidistributed as well, and the desired transformation $\mathbf{x} \Longleftrightarrow \mathbf{y}$ reads

$$\begin{pmatrix} x_1 \\ x_2 \end{pmatrix} \Longleftrightarrow \begin{pmatrix} e^{-(x_1^2 + x_2^2)/2} \\ \dfrac{1}{2\pi} \arctan \dfrac{x_2}{x_1} \end{pmatrix} \equiv \begin{pmatrix} y_1 \\ y_2 \end{pmatrix} \tag{3.35}$$

---

[2]A "cardboard and glue" method for producing almost normal variates makes use of the central limit theorem: If $y = x_1 + \ldots + x_n$ is the sum of $n = 10 - 15$ equidistributed random numbers picked from the interval $(-0.5, 0.5)$, then the distribution of $z \equiv y\sqrt{12/n}$ is almost normal.

---

**Box-Muller technique:**

- Sample $(y_1, y_2) \in (0,1)^2$

- Construct

$$
\begin{aligned}
x_1 &= \sqrt{-2 \ln y_1} \, \cos 2\pi y_2 \\
x_2 &= \sqrt{-2 \ln y_1} \, \sin 2\pi y_2
\end{aligned}
$$

The variables $x_1, x_2$ are then normal-distributed and statistically independent. Gaussian variates with given variances $\sigma_1^2, \sigma_2^2$ are obtained by multiplying $x_1$ and $x_2$ by their respective $\sigma_i$.

---

Figure 3.8: Gaussian random variates by the Box-Muller technique

Thus we may write up the Box-Muller prescription [MULLER 58] for generating normal random variates as shown in Figure 3.8. If one prefers to avoid the time-consuming evaluation of trigonometric functions, the method given in Section 3.2.6 may be used.

## 3.2.4    Rejection Method

The transformation method works fine only if the distribution function – i.e. the primitive function of the density – is known and invertible. What if $p(x)$ is too complicated for formal integration, or if it is given in tabulated form only, for instance as a measured angle-dependent scattering cross section? It was just this kind of problems the pioneers of stochastics had in mind when they taught ENIAC and MANIAC to play at dice. Therefore the classical method for generating arbitrarily distributed random numbers stems from those days. In a letter written by John von Neumann to Stanislaw Ulam in May 1947 we read:

> "An alternative, which works if $\xi$ and all values of $f(\xi)$ lie in 0, 1, is this: Scan pairs $x^i, y^i$ and use or reject $x^i, y^i$ according to whether $y^i \leq f(x^i)$ or not. In the first case, put $\xi^j = x^i$; in the second case form no $\xi^j$ at that step." [COOPER 89]

In Figure 3.9 this recipe is reproduced in modern notation. From Figure 3.10 it may be appreciated that by this prescription $x$-values with high $p(x)$ will indeed be accepted more frequently than others.

The method is simple and fast, but it becomes inefficient whenever the area of the rectangle $[a, b] \otimes [0, p_m]$ is large compared to the area below the graph of $p(x)$ (which by definition must be $= 1$). Therefore, if either the variation of $p(x)$ is large ("$\delta$-like $p(x)$") or the interval $[a, b]$ is extremely wide, a combination of transformation and rejection method is preferable. We first try to find a test

**Rejection method:**

Let $[a, b]$ be the allowed range of values of the variate $x$, and $p_m$ the maximum of the density $p(x)$.

1. Sample a pair of equidistributed random numbers, $x \in [a, b]$ and $y \in [0, p_m]$.

2. If $y \leq p(x)$, accept $x$ as the next random number, otherwise return to step 1.

Figure 3.9: Rejection method

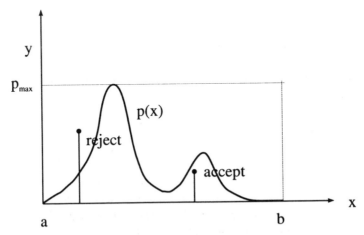

Figure 3.10: *Rejection method*

---

**Improved rejection method:**

Let $f(x)$ be a test function similar to $p(x)$, with

$$f(x) \geq p(x); \quad x \in [a, b] \tag{3.37}$$

The primitive function $F(x) \equiv \int f(x)\,dx$ is assumed to be known and invertible

1. Pick a random number $x \in [a, b]$ from a distribution with density

$$\bar{p}(x) = \frac{f(x)}{F(b) - F(a)} \tag{3.38}$$

by using the transformation method. Pick an additional random number $y$ equidistributed in the interval $[0, f(x)]$.

2. If $y \leq p(x)$ accept $x$ as the next random number, else return to Step 1.

---

Figure 3.11: Improved rejection method

function $f(x)$ which should closely resemble the desired density, with the additional requirement that $f(x) \geq p(x)$ everywhere. If $f(x)$ is integrable, with an invertible primitive $F'(x)$, we may employ the transformation method to generate $x$-values that are already distributed according to $f(x)$. More specifically, their distribution is given by the correctly normalized density

$$\bar{p}(x) \equiv \frac{f(x)}{F(b) - F(a)} \tag{3.36}$$

Now we pick a second random number $y$ from an equidistribution in $(0, f(x))$ and subject it to the test $y : p(x)$. By accepting $x$ only if $y \leq p(x)$ we generate $x$ with the correct distribution, but with more "hits" per trial than in the simple rejection technique (see Fig. 3.11).

The improvement with respect to the basic rejection method is related to the proximity of $f(x)$ to the given density $p(x)$. A test function that is particularly popular for use with single-peaked density functions is the Lorentzian introduced in equ. 3.30. The primitive of this function is known and invertible, which makes the first step in the improved rejection method very simple (see the example given in Section 3.2.2). Various applications of the improved method, all using this particular test function, may be found in the book by Press et al. [PRESS 86].

The rejection method will also be inefficient whenever $\mathbf{x} \equiv (x_1, \ldots x_n)$ is a high-dimensional vector. The probability that a sampled vector $\mathbf{x}$, in combination with $y \in (0, p_m)$, will be accepted according to the rule $y \leq p(\mathbf{x})$ is an $n$-fold product of probabilities and is therefore small. Multidimensional problems are better treated using a *random walk* (see Section 3.3.5). However, one must then accept that successive random vectors will not be uncorrelated.

There is one multidimensional distribution for which it is quite easy to generate random vectors. The following method for producing n-tuples of random numbers from a *multivariate Gaussian* distribution is formally elegant and works very fast.

## 3.2.5   Multivariate Gaussian Distribution

This is a – fortunately rather common – particular instance of a distribution of several random variates, $\mathbf{x} \equiv (x_1 \ldots x_n)$. Let us assume, for simplicity, that all individual averages are $\langle x_i \rangle = 0$. The density of the compound ("and") probability is given by

$$p(x_1, \ldots, x_n) = \frac{1}{\sqrt{(2\pi)^n\, S}}\, e^{-\frac{1}{2} \sum \sum g_{ij}\, x_i\, x_j} \tag{3.39}$$

or more concisely

$$p(\mathbf{x}) = \frac{1}{\sqrt{(2\pi)^n\, S}}\, e^{-\frac{1}{2}\mathbf{x}^T \cdot \mathbf{G} \cdot \mathbf{x}} \equiv \frac{1}{\sqrt{(2\pi)^n\, S}}\, e^{-\frac{1}{2} Q} \tag{3.40}$$

with the *covariance matrix* of the $x_i$

$$\mathbf{S} \equiv \mathbf{G}^{-1} \equiv \begin{pmatrix} \langle x_1^2 \rangle & \langle x_1\, x_2 \rangle & \cdots \\ \vdots & \langle x_2^2 \rangle & \cdots \\ & & \ddots \end{pmatrix} \tag{3.41}$$

$S \equiv |\mathbf{S}|$ is the determinant of this matrix. $\mathbf{S}$ and $\mathbf{G}$ are evidently symmetric, and as a rule they are diagonally dominated. Incidentally, we will obey custom by denoting the eigenvalues of the covariance matrix $\mathbf{S}$ by $\sigma_i^2$, while the eigenvalues of the inverse matrix $\mathbf{G}$ are simply called $\gamma_i$.

The quadratic form $Q \equiv \mathbf{x}^T \cdot \mathbf{G} \cdot \mathbf{x}$ describes a manifold ($Q = const$) of concentric n-dimensional ellipsoids whose axes will in general not coincide with the coordinate axes. If they do, then the matrices $\mathbf{S}$ and $\mathbf{G}$ are diagonal, and $p(\mathbf{x})$ decomposes into a product of $n$ independent probability densities:

$$p(\mathbf{x}) = \prod_{i=1}^{n} \frac{1}{\sqrt{2\pi\, s_{ii}}}\, e^{-\frac{1}{2} g_{ii} x_i^2} \tag{3.42}$$

Here $s_{ii} \equiv \langle x_i^2 \rangle$ and $g_{ii} = 1/s_{ii}$ are the diagonal elements of $\mathbf{S}$ and $\mathbf{G}$, respectively. (Besides, in this case $s_{ii} = \sigma_i^2$ and $g_{ii} = \gamma_i$, i.e. the diagonal elements are also the eigenvalues.) The $n$ variables $x_i$ are then uncorrelated and we may simply pick $n$ individual Gaussian variates, combining them to the vector $\mathbf{x}$.

EXAMPLE: Assume that two Gaussian variates have the variances $s_{11} \equiv \langle x_1^2 \rangle = 3$, $s_{22} \equiv \langle x_2^2 \rangle = 4$, and the covariance $s_{12} \equiv \langle x_1 x_2 \rangle = 2$:

$$\mathbf{S} = \begin{pmatrix} 3 & 2 \\ 2 & 4 \end{pmatrix}; \quad \mathbf{G} \equiv \mathbf{S}^{-1} = \begin{pmatrix} \frac{1}{2} & -\frac{1}{4} \\ -\frac{1}{4} & \frac{3}{8} \end{pmatrix}$$

The quadratic form $Q$ in the exponent of the probability density is then

$$Q = \frac{1}{2} x_1^2 - \frac{1}{2} x_1 x_2 + \frac{3}{8} x_2^2.$$

The lines of equal density (that is, of equal $Q$) are ellipses which are inclined with respect to the $x_{1,2}$ coordinate axes (see Fig. 3.12).

Incidentally, in this simple case one might generate the correlated random variates $x_1$, $x_2$ in the following manner:

- Draw $x_1$ from the *marginal* (also Gaussian) distribution

$$p(x_1) = \frac{1}{\sqrt{2\pi\, s_{11}}}\, e^{-\frac{1}{2 s_{11}} x_1^2} = \frac{1}{\sqrt{6\pi}}\, e^{-\frac{1}{6} x_1^2}$$

- Since $x_1$ is now fixed, $x_2$ may be picked from the *conditional* density (see 3.18)

$$p(x_2 | x_1) = \sqrt{\frac{s_{11}}{2\pi\, S}}\, e^{-\frac{s_{11}}{2 S}(x_2 - \frac{s_{12}}{s_{11}} x_1)^2} = \sqrt{\frac{3}{16\pi}}\, e^{-\frac{3}{16}(x_2 - \frac{2}{3} x_1)^2}$$

(This is the density of $x_2$ along the cut $x_1 = c$ in Fig. 3.12.)

For more than two correlated random variates this procedure is much too complicated. In contrast, the following method of principal axis transformation remains applicable for any number of dimensions.

If, in the foregoing example, the covariance had been $s_{12} \equiv \langle x_1 x_2 \rangle = 0$, we would have

$$p(x_1, x_2) = p(x_1)\, p(x_2) = \frac{1}{\sqrt{12\,(2\pi)^2}}\, e^{-\frac{1}{6} x_1^2 - \frac{1}{8} x_2^2}$$

All we would have to do is sample $x_1$ from a Gaussian distribution with $\sigma_1^2 = 3$ and $x_2$ with $\sigma_2^2 = 4$, then combine them to the vector $\mathbf{x} = (x_1, x_2)$. The ellipses $Q = const$ in Fig. 3.12 would have their axes parallel to the coordinate axes.

These considerations indicate a way to the production of correlated random numbers with the distribution density 3.39. If we could succeed in rotating the axes of the ellipsoids $Q = const$ by some linear transformation $\mathbf{x} = \mathbf{T} \cdot \mathbf{y}$ in such a way that they coincide with the coordinate axes, then $Q$ would be diagonal in terms of the new variables $(y_1 \ldots y_n)$. The transformed (y-) components of the vector $\mathbf{x}$ would be uncorrelated, and we could sample them independently.

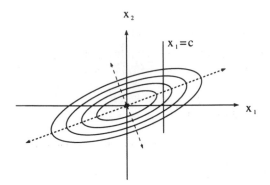

Figure 3.12: Bivariate Gaussian distribution: lines of equal density

What we have to find, then, is a transformation matrix $\mathbf{T}$ for which

$$Q = \mathbf{x}^T \cdot \mathbf{G} \cdot \mathbf{x} = \mathbf{y}^T \cdot \left[ \mathbf{T}^T \cdot \mathbf{G} \cdot \mathbf{T} \right] \cdot \mathbf{y} = \sum_{i=1}^{n} g_{ii}{}' y_i^2 \qquad (3.43)$$

where $g_{ii}{}'$ are the elements of the diagonalized matrix. This is an underdetermined problem, and we may choose among various possible diagonalization matrices $\mathbf{T}$. The generic method to construct a diagonalization matrix for a real, symmetric matrix $\mathbf{G}$ goes as follows:

---

**Principal axis transformation:**

- Determine the eigenvalues $\gamma_j$ and the eigenvectors $\mathbf{g}_j$ of $\mathbf{G}$. (There are standard subroutines available to perform this task, like NAG-F02AMF or ESSL-SSYGV.) If need be, normalize the $\mathbf{g}_j$ so that $|\mathbf{g}_j| = 1$.

- Combine the n column vectors $\mathbf{g}_j$ to form a matrix $\mathbf{T}$. This matrix diagonalizes $\mathbf{G}$ (and consequently the quadratic form $Q$.)

In this procedure, $\mathbf{S}$ may be used in place of $\mathbf{G} \equiv \mathbf{S}^{-1}$; the same diagonalization matrix $\mathbf{T}$ will result (see text).

---

As a special bonus the diagonalization matrix constructed in this manner is *orthogonal*, i.e. it has the property

$$\mathbf{T}^T = \mathbf{T}^{-1}. \qquad (3.44)$$

It follows that $\mathbf{T}$ diagonalizes not only $\mathbf{G} \equiv \mathbf{S}^{-1}$ but also the covariance matrix $\mathbf{S}$ itself:

$$\mathbf{T}^T \cdot \mathbf{S} \cdot \mathbf{T} = \mathbf{T}^{-1} \cdot \mathbf{S} \cdot \mathbf{T} = \left[ \mathbf{T}^{-1} \cdot \mathbf{S}^{-1} \cdot \mathbf{T} \right]^{-1} = \left[ \mathbf{T}^T \cdot \mathbf{G} \cdot \mathbf{T} \right]^{-1} \quad (\textit{diagonal}) \quad (3.45)$$

**Multivariate Gaussian distribution:**

Assume that the covariance matrix $S$ or its inverse $G$ is given. The matrix elements of $S$ are called $s_{ij}$, the eigenvalues are $\sigma_i^2$.

- Determine by the above method (principal axis transformation) the diagonalization matrix $T$ for $S$ or $G$. (This step is performed only once.)

- Generate $n$ mutually independent Gaussian random variates $y_i$ with the variances $\sigma_i^2$.

- Transform the vector $y \equiv (y_1 \ldots y_n)^T$ according to

$$x = T \cdot y \qquad (3.47)$$

The $n$ elements of the vector $x$ are then random numbers obeying the desired distribution 3.39.

Figure 3.13: Production of n-tuples of random numbers from a multivariate Gaussian distribution

This means that in in the above prescription for finding $T$ we may use $S$ instead of its inverse $G$, arriving at the same matrix $T$. For practical purposes, therefore, $G$ need not be known at all. All that is required are the covariances and the assumption that we are dealing with a multivariate Gaussian distribution.

Since $T$ is orthogonal and – by construction – unitary, we have for the diagonal elements of the transformed matrix $T^T \cdot G \cdot T$

$$g_{ii}' \equiv \gamma_i \equiv \frac{1}{\sigma_i^2} \qquad (3.46)$$

Thus we arrive at the prescription given in Figure 3.13 for the production of correlated Gaussian variables.

EXAMPLE: Once more, let

$$S = \begin{pmatrix} 3 & 2 \\ 2 & 4 \end{pmatrix}, \quad \text{with the inverse } G = \begin{pmatrix} \frac{1}{2} & -\frac{1}{4} \\ -\frac{1}{4} & \frac{3}{8} \end{pmatrix}$$

**Principal axis transformation:** The eigenvalues of $S$ are $\sigma_{1,2}^2 = (7 \pm \sqrt{17})/2 = 5.562|1.438$, and the corresponding eigenvectors are

$$s_1 = \begin{pmatrix} 0.615 \\ 0.788 \end{pmatrix} \quad s_2 = \begin{pmatrix} 0.788 \\ -0.615 \end{pmatrix}$$

(The eigenvectors of a real symmetric matrix are always mutually orthogonal.) The matrix constructed by combining $\mathbf{s}_1$ and $\mathbf{s}_2$,

$$\mathbf{T} = \begin{pmatrix} 0.615 & 0.788 \\ 0.788 & -0.615 \end{pmatrix}$$

should then diagonalize $\mathbf{S}$. We check this:

$$\begin{pmatrix} 0.615 & 0.788 \\ 0.788 & -0.615 \end{pmatrix} \cdot \begin{pmatrix} 3 & 2 \\ 2 & 4 \end{pmatrix} \cdot \begin{pmatrix} 0.615 & 0.788 \\ 0.788 & -0.615 \end{pmatrix} = \begin{pmatrix} 5.562 & 0 \\ 0 & 1.438 \end{pmatrix}$$

As stated above, the same matrix $\mathbf{T}$ will diagonalize the inverse $\mathbf{G}$ as well, and the remaining diagonal elements are simply the reciprocal values of the $\sigma_i^2$.

**Generator:** To produce a sequence of pairs $(x_1, x_2)$ of Gaussian random numbers with the given covariance matrix one has to repeatedly perform the following two steps:

- Draw $y_1$ and $y_2$ Gaussian, uncorrelated, with the variances 5.562 and 1.438, respectively. (For instance, one may sample two normal variates using the Box-Muller method and multiply them by $\sqrt{5.562}$ and $\sqrt{1.438}$, respectively.)

- Compute $x_1$ and $x_2$ according to

$$\begin{pmatrix} x_1 \\ x_2 \end{pmatrix} = \begin{pmatrix} 0.615 & 0.788 \\ 0.788 & -0.615 \end{pmatrix} \cdot \begin{pmatrix} y_1 \\ y_2 \end{pmatrix}$$

EXERCISE: Write a program that generates a sequence of bivariate Gaussian random numbers with the statistical properties as assumed in the foregoing example. Determine $\langle x_1^2 \rangle$, $\langle x_2^2 \rangle$, and $\langle x_1 x_2 \rangle$ to see if they indeed approach the given values of 3, 4, and 2.

## 3.2.6   Equidistribution in Orientation Space

Very often the radius vectors of points homogeneously distributed on the circumference of a circle are needed. To generate the cartesian coordinates of such points one could, of course, first sample an angle $\phi \in (0, 2\pi)$ and then compute $x_1 = r \cos \phi$ and $x_2 = r \sin \phi$. However, the evaluation of the two trigonometric functions is usually time-consuming and therefore undesirable. An alternative which need not be explained any further is given in Fig. 3.14. One has to discard a few random numbers (step 1) and evaluate a square root (step 2). However, the resulting expense in computer time is for most machines smaller than the gain achieved by avoiding the trigonometric functions.

It is worth mentioning that this technique may also be applied in the context of the Box-Muller method explained earlier, in order to avoid the evaluation of sine

---

**Equidistribution on the unit circle:**

- Draw a pair of equidistributed random numbers $(y_1, y_2) \in (-1, 1)^2$; compute $r^2 = y_1^2 + y_2^2$; if necessary, repeat until $r^2 \leq 1$.

- $x_1 \equiv y_1/r$ and $x_2 \equiv y_2/r$ are the cartesian coordinates of points that are homogeneously distributed on the circumference of the unit circle. (This means that we have generated cosine and sine of an angle $\phi$ equidistributed in $(0, 2\pi)$.)

---

Figure 3.14: Equidistribution on the circumference of a circle

---

**Marsaglia (3D):** To generate points homogeneously distributed on the surface of a sphere, proceed as follows:

- Draw pairs of random numbers $(y_1, y_2) \in (-1, 1)^2$ until $r^2 \equiv y_1^2 + y_2^2 \leq 1$.

- The quantities

$$
\begin{aligned}
x_1 &= 2y_1\sqrt{1 - r^2} \\
x_2 &= 2y_2\sqrt{1 - r^2} \\
x_3 &= 1 - 2r^2
\end{aligned}
$$

are then the cartesian coordinates of points out of a homogeneous distribution on the surface of the unit sphere.

---

Figure 3.15: Equidistribution on the surface of a sphere

and cosine. The first step is the same as in generating an equidistribution on the unit circle, while the second step in Fig. 3.14 is replaced by

$$
\begin{aligned}
x_1 &= y_1\sqrt{(-2\ln r^2)/r^2} & (3.48) \\
x_2 &= y_2\sqrt{(-2\ln r^2)/r^2} & (3.49)
\end{aligned}
$$

(Compare Fig. 3.8.)

The scheme given in Figure 3.14 may be generalized for higher dimensions [MARSAGLIA 72]). Thus, in case one needs points equidistributed over the surface of a sphere, one should not succumb to the temptation to introduce spherical polar coordinates, but should rather use the recipe of Figure 3.15. Somewhat more abstract, but still useful at times [VESELY 82] is the generalization to the 3-dimensional "surface" of a 4-dimensional unit sphere (see Figure 3.16).

---

**Marsaglia (4D):** To generate points equidistributed on the three-dimensional surface of a hypersphere:

- Draw pairs of random numbers $(y_1, y_2) \in (-1,1)^2$ until $r_1^2 \equiv y_1^2 + y_2^2 \le 1$.

- Draw pairs of random numbers $(y_3, y_4) \in (-1,1)^2$ until $r_2^2 \equiv y_3^2 + y_4^2 \le 1$.

- The quantities

$$
\begin{aligned}
x_1 &= y_1 \\
x_2 &= y_2 \\
x_3 &= y_3 \sqrt{(1 - r_1^2)/r_2^2} \\
x_4 &= y_4 \sqrt{(1 - r_1^2)/r_2^2}
\end{aligned}
$$

are then the cartesian coordinates of points out of a homogeneous distribution on the "surface" of a 4-dimensional unit sphere.

---

Figure 3.16: Equidistribution on the surface of a hypersphere

## 3.3   Random Sequences

### 3.3.1   Fundamentals

So far we have been concerned with the production of random numbers, which preferably should be free of serial correlations $\langle x_n x_{n+k} \rangle$. Next we will consider how to generate sequences of random numbers with *given* serial correlations. Once more we start out by reviewing a few basic concepts:

**Random process / random sequence:** Let $\{x(t)\}$ be an ensemble of functions of the time variable $t$. (Think of the set of all possible temperature curves in the course of a day, or the x-coordinate of a molecule in the course of its thermally agitated motion.) Once more we ascribe a probability distribution to the function values $x(t)$, which may vary within some given range $(a, b)$:

$$P_1(x; t) \equiv \mathcal{P}\left\{x(t) \le x\right\} \tag{3.50}$$

By the same token a probability density

$$p_1(x; t) \equiv \frac{dP_1(x; t)}{dx} \tag{3.51}$$

is defined. Such an ensemble of time functions is called a random process. A particular function $x(t)$ from the ensemble is called a *realization* of the random process.

A random process is called a *random sequence* if the variable $t$ may assume only discrete values $\{t_k \, ; \; k = 0, 1, \ldots \}$. In this case one often writes $x(k)$ for $x(t_k)$.

<u>EXAMPLE:</u> Let $x_0(t)$ be a deterministic function of time, and assume that the quantity $x(t)$ at any time $t$ be Gauss distributed about the value $x_0(t)$:

$$p_1(x;t) = \frac{1}{\sqrt{2\pi\sigma^2}} \, e^{-\frac{1}{2} \, [x - x_0(t)]^2 / \sigma^2}$$

(Of course the variance $\sigma$ might be a function of time as well.)

**Distribution functions of higher order:** The foregoing definitions may be generalized in the following manner:

$$P_2(x_1, x_2; t_1, t_2) \;\equiv\; \mathcal{P}\{x(t_1) \le x_1, x(t_2) \le x_2\} \qquad (3.52)$$

$$\vdots$$

$$P_n(x_1, \ldots, x_n; t_1, \ldots, t_n) \;\equiv\; \mathcal{P}\{x(t_1) \le x_1, \ldots, x(t_n) \le x_n\}$$
$$(3.53)$$

Thus $P_2(..)$ is the compound probability for the events $x(t_1) \le x_1$ *and* $x(t_2) \le x_2$. These higher order distribution functions and the corresponding densities

$$p_n(x_1, \ldots, x_n; t_1, \ldots, t_n) = \frac{d^n P_n(x_1, \ldots, x_n; t_1, \ldots, t_n)}{dx_1 \ldots dx_n} \qquad (3.54)$$

describe the random process in ever more – statistical – detail.

**Stationarity:** A random process is stationary in the strong sense if for all higher distribution functions

$$P_n(x_1, \ldots, x_n; t_1, \ldots, t_n) = P_n(x_1, \ldots, x_n; t_1 + t, \ldots, t_n + t), \qquad (3.55)$$

This means that the origin of time is of no importance. The functions $P_1(x;t)$ and $p_1(x;t)$ are then not dependent upon time at all: $P_1(x;t) = P_1(x)$, $p_1(x;t) = p_1(x)$. Furthermore, $P_2(\ldots)$ and $p_2(\ldots)$ depend only on the time difference $\tau \equiv t_2 - t_1$:

$$p_2(x_1, x_2; t_1, t_2) = p_2(x_1, x_2; \tau). \qquad (3.56)$$

A random process is stationary of order $k$ if the foregoing condition is fulfilled for the distribution functions up to $k$-th order only. In the following we will treat only random processes that are stationary of second order.

**Moments:** The moments of the distribution density 3.51 are defined in the same way as for simple random variates:

$$\langle x^n(t) \rangle \equiv \int_a^b x^n \, p_1(x;t) \, dx \tag{3.57}$$

(In the stationary case this is indeed identical to the definition 3.20.) In addition we may now define moments of the distribution density of second order (viz. 3.56):

$$\langle x^m(t_1) \, x^n(t_2) \rangle \equiv \int_a^b \int_a^b x_1^m x_2^n \, p_2(x_1, x_2; t_1, t_2) \, dx_1 \, dx_2 \tag{3.58}$$

In the stationary case things depend on the temporal distance $\tau \equiv t_2 - t_1$ only:

$$\langle x^m(0) \, x^n(\tau) \rangle \equiv \int_a^b \int_a^b x_1^m x_2^n \, p_2(x_1, x_2; \tau) \, dx_1 \, dx_2 \tag{3.59}$$

**Autocorrelation:** A particularly important moment of the second order density is the quantity

$$\langle x(0) \, x(\tau) \rangle \equiv \int_a^b \int_a^b x_1 x_2 \, p_2(x_1, x_2; \tau) \, dx_1 \, dx_2 \,, \tag{3.60}$$

which is called the *autocorrelation function* of $x(t)$. For $\tau \to 0$ it approaches the variance $\langle x^2 \rangle$. For finite $\tau$ it tells us how rapidly a particular value of $x(t)$ will be "forgotten". To see this we may make use of the *conditional* density (viz. equ. 3.18):

$$p(x_2|x_1; \tau) = \frac{p_2(x_1, x_2; \tau)}{p_1(x_1)} \tag{3.61}$$

is the density of $x_2$ at time $t + \tau$ *under the condition* that at time $t$ we had $x(t) = x_1$. The *conditional moment*

$$\langle x(\tau) \,|\, x_1 \rangle \equiv \int x_2 \, p(x_2|x_1; \tau) \, dx_2 \tag{3.62}$$

is then the average of $x(t+\tau)$ under the same condition. The faster $p(x_2|x_1; \tau)$ decays with $\tau$ the more rapidly the conditional average will approach the unconditional one:

$$\langle x(\tau) \,|\, x_1 \rangle \;\to\; \langle x \rangle \tag{3.63}$$

For later reference we note that the definition 3.61 may be generalized as

$$p(x_n|x_{n-1}, \ldots x_1; t_n, \ldots t_1) = \frac{p_n(x_1, \ldots x_n; t_1, \ldots t_n)}{p_{n-1}(x_1, \ldots x_{n-1}; t_1, \ldots t_{n-1})} \tag{3.64}$$

**Gaussian process:** A random process is a (stationary) Gaussian process if the
random variables $x(t_1), \ldots, x(t_n)$ obey a multivariate Gaussian distribution.
The matrix elements of the covariance matrix – which, as we know, determines the distribution uniquely (see Section 3.2.5) – are in this case simply
the values of the autocorrelation function at the respective time displacements, $\langle x(0)\, x(t_j - t_i)\rangle$. A Gauss process, then, is uniquely determined by its
autocorrelation function; the distribution function is just

$$p_1(x) = \frac{1}{\sqrt{2\pi\sigma^2}}\, e^{-\frac{1}{2}\,x^2/\sigma^2} \tag{3.65}$$

with $\sigma^2 \equiv \langle x^2 \rangle$. Furthermore we have

$$p_2(x_1, x_2\,;\tau) = \frac{1}{\sqrt{(2\pi)^2 S_2(\tau)}}\, e^{-\frac{1}{2}\,Q} \tag{3.66}$$

with

$$Q \equiv \frac{\langle x^2 \rangle x_1^2 - 2\langle x(0)x(\tau)\rangle x_1\,x_2 + \langle x^2 \rangle x_2^2}{S_2(\tau)} \tag{3.67}$$

and

$$S_2(\tau) \equiv |\mathbf{S}_2(\tau)| = \langle x^2 \rangle^2 - \langle x(0)\,x(\tau)\rangle^2 \tag{3.68}$$

Similarly,

$$p_n(x_1 \ldots x_n\,; t_1 \ldots t_n) = \frac{1}{\sqrt{(2\pi)^n S_n}}\, e^{-\frac{1}{2}\mathbf{x}^T \cdot \mathbf{S}_n^{-1} \cdot \mathbf{x}} \tag{3.69}$$

where the elements of $\mathbf{S}$ are simply given by $\langle x(t_i)\,x(t_j)\rangle$, which in the stationary case is identical to $\langle x(0)\,x(t_j - t_i)\rangle$.

## 3.3.2   Markov Processes

For the sake of simplicity we will restrict the discussion to random *sequences*, i.e.
random processes on a discretized time axis. A stationary random sequence is said
to have the *Markov property* if

$$p(x_n|x_{n-1} \ldots x_1) = p(x_n|x_{n-1}) \tag{3.70}$$

Thus it is assumed that the "memory" of the physical system we try to model
by the random sequence goes back no farther than to the preceding step. All
elements of the sequence ($\hat{=}$ "states" of the model system) that are farther back do
not influence the distribution density of the n-th element. An even shorter memory
would mean that successive elements of the sequence were not correlated at all.

   Of particular practical importance are *Gaussian Markov* processes. To describe
them uniquely not even $p_2(\ldots)$ is needed. It is sufficient that the autocorrelation
function $\langle x(0)\,x(\tau)\rangle$ be known; then $p_2(..)$ and consequently all statistical properties of the process follow. Incidentally, it is an important hallmark of stationary

Gaussian Markov processes that their autocorrelation function is always an exponential:

$$\langle x(0)\, x(\tau) \rangle = \langle x^2 \rangle e^{-\beta \tau} \tag{3.71}$$

For a proof see [PAPOULIS 81].

The most simple procedure for generating a stationary Gaussian Markov process is based on the stepwise solution of the stochastic differential equation

$$\dot{x}(t) = -\beta\, x(t) + s(t) \tag{3.72}$$

with a stochastic "driving" process $s(t)$. For some given $x(0)$ the general solution to this equation reads

$$x(t) = x(0)\, e^{-\beta t} + \int_0^t e^{-\beta(t-t')}\, s(t')\, dt' \tag{3.73}$$

Inserting $t = t_n$ and $t = t_{n+1} \equiv t_n + \Delta t$ one finds that

$$x(t_{n+1}) = x(t_n)\, e^{-\beta\, \Delta t} + \int_0^{\Delta t} e^{-\beta(\Delta t - t')}\, s(t_n + t')\, dt' \tag{3.74}$$

The equation of motion 3.72 is complete only if the statistical properties of $s(t)$ are given as well. We will assume that $s(t)$ be Gauss distributed about $\langle s \rangle = 0$, with

$$\langle s(0)\, s(t) \rangle = A\, \delta(t) \tag{3.75}$$

The driving random process is thus assumed to be uncorrelated noise. (This is often called "$\delta$-correlated noise".) With these simple assumptions it may be shown that the values of the solution function $x(t)$ (equ. 3.73) at any time $t$ belong to a stationary Gaussian distribution with $\langle x^2 \rangle = A/2\beta$ and that the process $\{x(t_n)\}$ has the Markov property.

To obtain a prescription for producing the stepwise solution 3.74 we interpret the integrals

$$z(t_n) \equiv \int_0^{\Delta t} e^{-\beta(\Delta t - t')}\, s(t_n + t')\, dt' \tag{3.76}$$

as elements of a random sequence whose statistical properties may be derived from those of the quantity $s(t)$. In particular, $z$ is Gauss distributed with zero mean and $\langle z(t_n)\, z(t_{n+k}) \rangle = 0$ for $k \neq 0$. The variance is

$$\langle z^2 \rangle = \frac{A}{2\beta}\left(1 - e^{-2\beta\, \Delta t}\right) \tag{3.77}$$

From all this there follows the recipe given in Figure 3.17 for generating a stationary, Gaussian Markov sequence.

EXAMPLE: Consider one cartesian component $v(t)$ of the velocity of a massive molecule

---

**"Langevin Shuffle":**

Let the desired stationary Gaussian Markov sequence $\{x(n)\,;\ n = 0,\ldots\}$ be defined by the autocorrelation function

$$\langle x(n)\,x(n+k)\rangle = \frac{A}{2\beta}\,e^{-\beta\,k\,\Delta t} \qquad (3.78)$$

with given parameters $A$, $\beta$ and $\Delta t$. A starting value $x(0)$ is chosen, either by putting $x(0) = 0$ or by sampling $x(0)$ from a Gauss distribution with $\langle x \rangle = 0$ and $\langle x^2 \rangle = A/2\beta$.

- Draw $z(n)$ from a Gaussian distribution with $\langle z \rangle = 0$ and

$$\langle z^2 \rangle = \frac{A}{2\beta}\,(1 - e^{-2\beta\,\Delta t}) \qquad (3.79)$$

- Construct

$$x(n+1) = x(n)\,e^{-\beta\,\Delta t} + z(n) \qquad (3.80)$$

The random sequence thus produced has the desired properties.

If the product $\beta\,\Delta t$ is much smaller than 1, the exponential in the foregoing formulae may be replaced by the linear Taylor approximation. The iteration prescription then reads

$$x(n+1) = x(n)\,(1 - \beta\,\Delta t) + z'(n) \qquad (3.81)$$

where $z'(n)$ is picked from a Gauss distribution with $\langle z'^2 \rangle = A\,\Delta t\,(1 - \beta\,\Delta t)$.

---

Figure 3.17: Generating a stationary Gaussian Markov sequence

undergoing diffusive motion in a solvent. It is a fundamental truth of statistical mechanics that this quantity is Gauss distributed with variance $kT/m$:

$$p_1(v;t) = p_1(v) = \frac{1}{\sqrt{2\pi(kT/m)}} e^{-\frac{mv^2}{2kT}}$$

Furthermore, under certain simplifying assumptions one may show that the random process $v(t)$ obeys the equation of motion postulated by Paul Langevin,

$$\dot{v}(t) = -\beta\, v(t) + s(t) \tag{3.82}$$

Here $\beta$ is a friction coefficient, and the stochastic acceleration $s(t)$ is a $\delta$−correlated Gaussian process with the autocorrelation function $\langle s(0)\, s(t)\rangle = (2\beta\, kT/m)\, \delta(t)$.

Again introducing a finite time step $\Delta t$ we can generate a realization of the random process $v(t)$ by the method explained above. In this case we have $A = 2\beta\, kT/m$, which means that the uncorrelated random variate $z(n)$ must be sampled from a Gauss distribution with $\langle z^2\rangle = (kT/m)(1 - exp(-2\beta\,\Delta t))$.

The process $v(t)$ as described by 3.82 is stationary and Gaussian with the autocorrelation function

$$\langle v(0)\, v(\tau)\rangle = \frac{kT}{m}\, e^{-\beta\tau} \tag{3.83}$$

By some further analysis we could obtain the position $x(t)$ as well, in addition to the velocity. This method of simulating the random motion of a dissolved particle is called "Stochastic dynamics" or "Brownian dynamics". It will be reviewed at more length in Chapter 6.

EXERCISE: Employ the procedure 3.80 to generate a Markov sequence $\{x_n\}$ and check if its autocorrelation function indeed has the form 3.78.

## 3.3.3   Autoregressive Processes

We have seen that an iterative procedure of the form

$$x(n+1) = a\, x(n) + z(n), \tag{3.84}$$

with Gaussian $z(n)$ will automatically produce a Gaussian Markov process. The Markov property – the "forgetfulness" of the system – is expressed by the fact that the distribution of $x(n+1)$ depends on the value of $x(n)$ only.

A natural generalization of this prescription reads

$$x(n+1) = \sum_{k=1}^{K} a_k\, x(n+1-k) + z(n) \tag{3.85}$$

where $z(n)$ is again a $\delta$-correlated process that is not correlated with $x(n)$ or any of the foregoing $x(n-m)$:

$$\langle x(n+1-k)\, z(n)\rangle = 0; \quad k = 1, 2, \ldots \tag{3.86}$$

Equation 3.85 describes a process in which earlier members of the sequence exert some influence on the probability density of $x(n+1)$. Thus the coefficients $a_k$ are table values of a "memory function" describing the effect of past states on $x(n+1)$.[3] In the case of the simple Markov sequence we have $a_k = a\,\delta_{k1}$.

Normally the table $\{a_k;\ k=1,\ldots,K\}$ will not be given a priori. Rather, the random sequence will be known (or required) to have a certain autocorrelation function:

$$c_m \equiv \langle x(n)\,x(n+m)\rangle\ ;\quad m = 0,1,\ldots \tag{3.87}$$

How, then, can one determine the coefficients $a_k$ such that they produce, when inserted in 3.85, a random sequence with the desired autocorrelation?

Let us assume that the autocorrelation function (ACF, from now on) be negligible after $M$ steps: $c_m \approx 0$ for $m > M$. Now multiply each of the $M$ equations

$$x(n+m) = \sum_{k=1}^{K} a_k\, x(n+m-k) + z(n+m-1)\ ;\quad m = 1,\ldots,M \tag{3.88}$$

by $x(n)$ and take the average to find

$$c_m = \sum_{k=1}^{K} a_k\, c_{m-k}\ ;\quad m = 1,\ldots,M \tag{3.89}$$

In matrix notation this reads

$$\mathbf{c} = \mathbf{C}\cdot\mathbf{a} \tag{3.90}$$

with $\mathbf{c}\equiv\{c_1,\ldots,c_M\}$, $\mathbf{a}\equiv\{a_1,\ldots,a_K\}$, and

$$\mathbf{C} = \begin{pmatrix} c_0 & c_1 & \cdot & \cdot & c_{K-1} \\ c_1 & c_0 & c_1 & \cdot & c_{K-2} \\ c_2 & & \cdot & & \cdot \\ \cdot & & & \cdot & \cdot \\ c_{M-1} & \cdot & \cdot & \cdot & c_{M-K} \end{pmatrix} \tag{3.91}$$

Here we have taken into account that the ACF of a stationary process is a symmetric function of time: $c_{-m} = c_m$. In communication science the $M$ equations 3.89 and 3.90 with the $K$ unknowns $a_k$ are known as *Yule-Walker equations* [HONERKAMP 91].

In most cases far less than $M$ table values $a_k$ ($k = 1,\ldots K$) are needed to generate an ACF given by $M$ values. For example, in the case of a simple Markov sequence the instantly decaying memory function $a_k = a\,\delta_{k1}$ already produces an exponentially, i.e. less rapidly, decaying ACF. However, for $K < M$ the system of equations 3.90 is overdetermined, and we cannot fulfill it exactly. In such cases one attempts to optimize the $a_k$ in such a way that the desired ACF is at least approximately reproduced. The approximation error consists of the elements $\varepsilon_m \equiv$

---

[3] The exact definition of the *memory function* will be given in Section 6.6.

$c_m - \sum_{k=1}^{K} a_k c_{m-k}$, and we will try to minimize the quantity $\sum_{m=1}^{M} \varepsilon_m^2$. This leads us to the equations

$$\mathbf{C}^T \cdot \mathbf{C} \cdot \mathbf{a} = \mathbf{C}^T \cdot \mathbf{c} \qquad (3.92)$$

Having determined the coefficients $a_k$, we use the relation

$$\langle z^2 \rangle = c_0 - \sum_{k=1}^{K} a_k c_k \qquad (3.93)$$

to calculate that variance of the random process $z(n)$ which is needed to produce, by applying 3.85, a random sequence $\{x(n)\}$ with the desired properties [SMITH 90, NILSSON 90].

EXAMPLE: The desired ACF is given as $c_0 = 1$, $c_1 = 0.9$, $c_2 = 0.5$, $c_3 = 0.1$. We want to find an autoregressive process of order $K = 2$ whose ACF approximates the given table $\{c_m, \ m = 0, \ldots, 3\}$. The matrix $\mathbf{C}$ is given by

$$\mathbf{C} = \begin{pmatrix} 1.0 & 0.9 \\ 0.9 & 1 \\ 0.5 & 0.9 \end{pmatrix} \qquad (3.94)$$

and equation 3.92 reads

$$\begin{pmatrix} 2.06 & 2.25 \\ 2.25 & 2.62 \end{pmatrix} \cdot \begin{pmatrix} a_1 \\ a_2 \end{pmatrix} = \begin{pmatrix} 1.0 & 0.9 & 0.5 \\ 0.9 & 1 & 0.9 \end{pmatrix} \cdot \begin{pmatrix} 0.9 \\ 0.5 \\ 0.1 \end{pmatrix} \qquad (3.95)$$

The solution is

$$\mathbf{a} = \begin{pmatrix} 1.55 \\ -0.80 \end{pmatrix} \qquad (3.96)$$

Let us check whether this process indeed has an ACF that fits the given $c_m$-values: $c_0 a_1 + c_1 a_2 = 0.83$ (instead of 0.9), $c_1 a_1 + c_0 a_2 = 0.60$ (for 0.5), $c_2 a_1 + c_1 a_2 = 0.06$ (for 0.1).

The correct variance $\langle x^2 \rangle = c_0$ is obtained by choosing for $\langle z^2 \rangle$ the value $c_0 - a_1 c_1 - a_2 c_2 = 0.005$ (see equ. 3.93).

EXERCISE: Write a program to generate a random sequence with the ACF given above. Test the code by computing the ACF of the sequence thus produced.

When trying to invert the matrix $\mathbf{C}^T \cdot \mathbf{C}$ one may run into trouble. Quite generally, fitting problems of this kind often lead to almost singular matrices. There are well-proven ways to deal with such situations, and "Numerical Recipes" by PRESS et al. is again a good source to turn to for help [PRESS 86].

To make an ad hoc suggestion: One may solve – uniquely – the first $K$ equations of the overdetermined system 3.90. Then the values $\{a_k, \ k = 1, \ldots K\}$ may be used as initial estimates in an iterative procedure treating the full system (see Sec. 2.2). (However, we then have to expect a rather low convergence rate.)

---

**Wiener-Lévy process:**

Let $A$ and $\Delta t$ (or just the product $A\Delta t$) be given. Choose $x(0) = 0$.

- Pick $z(n)$ from a Gauss distribution with zero mean and variance $A\Delta t$.

- Compute

$$x(n+1) = x(n) + z(n) \qquad (3.100)$$

The random sequence thus produced is a nonstationary Gaussian process with variance $[x(n)]^2 = n\,A\,\Delta t$.

---

Figure 3.18: Unbiased random walk

## 3.3.4   Wiener-Lévy Process

Consider once more the stochastic differential equation 3.72. If we take the parameter $\beta$ to be zero, the $x$-increment for the step $t_n \to t_n + \Delta t$ equals (see equ. 3.80)

$$x(n+1) = x(n) + z(n) \qquad (3.97)$$

where

$$z(n) \equiv \int_0^{\Delta t} s(t_n + t')\,dt' \qquad (3.98)$$

is a Gaussian random variate with $\langle z \rangle = 0$ and $\langle z^2 \rangle = A\Delta t$. Since $z$ and $x$ are uncorrelated, we have

$$\langle [x(n)]^2 \rangle = n\,A\,\Delta t \qquad (3.99)$$

Thus the variance of $x$ now increases linearly with the number of steps. In other words, this random process is no more stationary.

As an example, interpreting $x$ as one cartesian coordinate of a diffusing particle we identify $\langle [x(n)]^2 \rangle$ with the mean squared displacement after $n$ time steps. In this case we may relate the coefficient $A$ to the diffusion constant according to $A = 2D$.

A stochastic process obeying equ. 3.98 is called a *Wiener-Lévy* process, or *Brownian (unbiased) random walk* (see Fig. 3.18).

EXERCISE: 500 *random walkers* set out from positions $x(0)$ homogeneously distributed in the interval $[-1, 1]$. The initial particle density is thus rectangular. Each of the random walkers is now set on its course to perform its own one-dimensional trajectory according to equ. 3.100, with $A\,\Delta t = 0.01$. Sketch the particle density after 100, 200, ... steps.

Incidentally, it is not really necessary to draw $z(n)$ from a Gaussian distribution. For instance, if $z(n)$ comes from an equidistribution in $[-\Delta x/2, \Delta x/2]$, the central limit theorem will enforce that the "compound" $x$-increment after every $10 - 15$

steps will again be Gauss distributed. (See the footnote on page 58.) We may even discretize the $x$-axis and allow single steps of the form $z = 0$, $+\Delta x$ or $-\Delta x$ only, with equal probability $1/3$ for any of these. After many steps, and on a scale which makes $\Delta x$ appear small, the results will again be the same as before.

To simulate a 2- or 3-dimensional diffusion process one simply applies the above procedure simultaneously and independently to 2 or 3 particle coordinates.

### 3.3.5   Markov Chains and the Monte Carlo method

A Markov sequence in which the variable $x_\alpha$ can assume discrete values only is called a *Markov chain*. As there is no reason to restrict the discussion to scalar variables, we will consider a discrete set of "state vectors" $\{\mathbf{x}_\alpha, \alpha = 1, \dots M\}$. The conditional probability

$$p_{\alpha\beta} \equiv \mathcal{P}\left\{\mathbf{x}(n) = \mathbf{x}_\beta \,|\, \mathbf{x}(n-1) = \mathbf{x}_\alpha\right\} \tag{3.101}$$

is then called *transition probability* between the states $\alpha$ and $\beta$.

Let $M$ be the total number – not necessarily finite – of possible states. The $M \times M$-matrix $\mathbf{P} \equiv \{p_{\alpha\beta}\}$ and the $M$-vector $\mathbf{p}$ consisting of the individual probabilities $p_\alpha \equiv \mathcal{P}\{\mathbf{x} = \mathbf{x}_\alpha\}$ determine the statistical properties of the Markov chain uniquely.

We are dealing with a *reversible* Markov chain if

$$p_\alpha \, p_{\alpha\beta} = p_\beta \, p_{\beta\alpha} \tag{3.102}$$

Recalling that $p_\alpha \, p_{\alpha\beta}$ is the probability that at some step (the $n$-th, say) the state $\mathbf{x} = \mathbf{x}_\alpha$ is realized and that at the next step we have $\mathbf{x} = \mathbf{x}_\beta$, the property of reversibility simply means that the same combined event in reverse order (i.e. $\mathbf{x} = \mathbf{x}_\beta$ at step $n$ and $\mathbf{x} = \mathbf{x}_\alpha$ at step $n+1$) is equally probable.

The $M^2$ elements of the matrix $\mathbf{P}$ are not uniquely defined by the $M(M-1)/2$ equations 3.102. For a given distribution density $\mathbf{p}$ we therefore have the choice between many possible transition matrices fulfilling the reversibility condition. A particularly popular recipe is the so-called "asymmetrical rule" introduced by N. Metropolis:

Assume that all $\mathbf{x}_\beta$ within a certain region around $\mathbf{x}_\alpha$ may be reached with the same a priori probability $\pi_{\alpha\beta} = 1/Z$, where $Z$ denotes the number of these $\mathbf{x}_\beta$ (including $\mathbf{x}_\alpha$ itself.) We then set the rule

$$p_{\alpha\beta} \;=\; \pi_{\alpha\beta} \qquad\qquad \text{if } p_\beta \geq p_\alpha \tag{3.103}$$

$$p_{\alpha\beta} \;=\; \pi_{\alpha\beta} \, \frac{p_\beta}{p_\alpha} \qquad \text{if } p_\beta < p_\alpha \tag{3.104}$$

It is easy to see that this rule fulfills the reversibility condition 3.102. Another widely used prescription is the *symmetrical*, or *Glauber*, rule

$$p_{\alpha\beta} = \pi_{\alpha\beta} \, \frac{p_\beta}{p_\alpha + p_\beta} \tag{3.105}$$

**Random numbers à la Metropolis:**

Let $\mathbf{p} \equiv \{p_\alpha; \ \alpha = 1, 2, \ldots\}$ be the vector of probabilities of the events $x = x_\alpha$. We want to generate a random sequence $\{x(n)\}$ in which the relative frequency of the event $x(n) = x_\alpha$ approaches $p_\alpha$.

- After the $n$-th step, let $x(n) = x_\alpha$. Draw a value $x_\beta$ from a region around $x_\alpha$, preferably according to

$$x_\beta = x_\alpha + (\xi - 0.5)\, \Delta x$$

  where $\xi$ is a random number from an equidistribution $\in (0,1)$, and where $\Delta x$ defines the range of directly accessible states $x_\beta$. (This recipe corresponds to the a priori transition probability $\pi_{\alpha\beta} = 1/Z$; note, however, that other symmetric a priori probabilities are permissible.)

- If for $p_\beta \equiv p(x_\beta)$ we have $p_\beta \geq p_\alpha$, then let $x(n+1) = x_\beta$.

- If $p_\beta < p_\alpha$, then pick a random number $\xi$ from an equidistribution $\in (0,1)$; if $\xi < p_\beta/p_\alpha$, let $x(n+1) = x_\beta$; else put $x(n+1) = x_\alpha$.

It is recommended to adjust the parameter $\Delta x$ such that approximately one out of two trial moves leads to a new state, $x(n+1) = x_\beta$.

Figure 3.19: Random numbers by a *biased random walk*

(Incidentally, other a priori transition probabilities than $1/Z$ may be used; all that is really required is that they are symmetrical with respect to $\alpha$ and $\beta$.)

Now for the important point. There is a beautiful theorem on reversible stationary Markov chains which in fact may be regarded as the central theorem of the Monte Carlo method (see Chapter 6):

> If the stationary Markov chain characterized by $\mathbf{p} \equiv \{p_\alpha\}$ and $\mathbf{P} \equiv \{p_{\alpha\beta}\}$ is reversible, then each state $\mathbf{x}_\alpha$ will be visited, in the course of a sufficiently long chain, with the relative frequency $p_\alpha$.

We may utilize this theorem together with the asymmetric or symmetric rule to formulate yet another recipe for generating random numbers with a given probability density $\mathbf{p}$. This procedure is described in Figure 3.19. It is also sometimes called a *random walk*, and to discern it from the Wiener-Lévy process the name *biased random walk* is often preferred. Recall that in a simple (unbiased) random walk on the discretized $x$-axis the transition probability to all possible neighboring positions is symmetric about $x(n) = x_\alpha$. (In the most simple procedure only the positions $x_{\alpha\pm1}$ or $x_\alpha$ are permitted as the new position $x(n+1)$, and the probabilities for $x_{\alpha+1}$ and $x_{\alpha-1}$ are equal.)

Thus the method of the *biased random walk* generates random numbers with the required distribution. However, in contrast to the techniques discussed in Section 3.2 this method produces random numbers that are serially correlated: $\langle x(n)\, x(n+k)\rangle \neq 0$.

EXERCISE: Serial correlations among pseudorandom numbers are normally regarded as undesirable, and the use of the biased random walk for a random number generator is accordingly uncommon. In spite of this we may test the method using a simple example. Let $p(x) = A\, exp[-x^2]$ be the desired probability density. Apply the prescription given in Fig. 3.19 to generate random numbers with this density. Confirm that $\langle x(n)\, x(n+k)\rangle \neq 0$.

An essential advantage of this method should be mentioned which more than makes up for the inconvenient serial correlations. In the transition rules, symmetric or asymmetric, the probabilities of the individual states appear only in terms of ratios $p_\beta/p_\alpha$ or $p_\beta/(p_\alpha + p_\beta)$. This means that their absolute values need not be known at all! Accordingly, in the preceding exercise the normalizing factor of $p_\alpha$, which we simply called $A$, never had to be evaluated.

In the most prominent application of the biased random walk, namely the statistical-mechanical Monte Carlo simulation, the state vector $\mathbf{x}_\alpha$ is a configuration vector comprised of $3N$ coordinates, with $N$ the number of particles in the model system. The probability $p_\alpha$ is there given by the thermodynamic probability of a configuration. As a rule we do not know this probability in absolute terms. We only know the Boltzmann factor which is indeed proportional to the probability, but with a usually inaccessible normalizing factor, the *partition function*.

Thus the feasibility of the Monte Carlo technique hinges on the fact that in a biased random walk the probabilities of the individual states need be known only up to some normalizing factor. The above theorem guarantees that in a correctly performed random walk through $3N$-dimensional configuration space all possible positions of the $N$ particles will be realized with their appropriate relative frequencies (see Sec. 6.2).

## 3.4   Stochastic Optimization

Optimization problems pop up in many branches of applied mathematics. They may always be interpreted as the task of finding the global extremum of a function of many variables. Examples are the nonlinear fit to a given set of table values (the function to be minimized being the sum of squared deviations), the improvement of complex electronic circuits ("travelling salesman problem"), or finding the most stable (i. e. lowest energy) configuration of microclusters or biopolymers.

A *systematic* scan of variable space for such a global extremum is feasible only for up to $6-8$ variables. Above that, a simple *stochastic* method would be to repeatedly draw a starting position and find the nearest local minimum by a steepest descent strategy. However, if the function to be minimized has a very ragged profile, this procedure will again be slow in identifying the lowest one among all local minima.

Thus it came as a welcome surprise that there is a much more efficient stochastic method to detect the global extremum of a function of many variables. In the Eighties, Kirkpatrick et al. [KIRKPATRICK 83] found that the Monte Carlo principle introduced in 3.3.5 may be employed in this task. Since the principle of the method resembles a cautious cooling of a thermodynamic system, the technique came to be called "Simulated Annealing".

Yet another group of optimization methods had been developed even earlier. Called evolutionary algorithms (EA) or genetic algorithms (GA), they are nowadays applied to optimization tasks in such widely separated fields as material science, biochemistry, artificial intelligence, and commerce.

## 3.4.1   Simulated Annealing

When performing a Monte Carlo walk through the set of possible events, or "states" $\mathbf{x}_\alpha$, following Metropolis' directions, we occasionally penetrate into regions of smaller probability $p_\alpha$. Let us now write this probability as

$$p_\alpha = A \exp -\beta U(\mathbf{x}) \tag{3.106}$$

where $U(x_1, \ldots x_M)$ is a "cost function" to be minimized, and $\beta$ a tunable parameter. The lower the value of $\beta$, the smaller the variation of the probabilities $p_\alpha$. Referring to the procedure given in Fig. 3.19 we see that it is then easy to visit the "high ranges" of the $U(\mathbf{x})$ landscape. As the parameter $\beta$ is increased, the point $\mathbf{x}$ representing the state of our system will preferably move "downhill". Eventually, for $\beta \to \infty$ only the nearest local minimum of the function $U(\mathbf{x})$ can be reached at all.

The probability 3.106 closely resembles the Boltzmann factor of statistical thermodynamics, giving the probability of a configuration $\mathbf{x}$ having energy $U(\mathbf{x})$. Accordingly, we may interpret the parameter $\beta$ as a reciprocal temperature, $\beta \equiv 1/kT$. High temperatures then refer to high accessibility of all regions of configuration space, and by lowering the temperature $kT$ we gradually force the system to remain in regions where $U(\mathbf{x})$ is low. In material technology such slow cooling is called *annealing*, which explains the name "Simulated Annealing" for the present method.

In practice one proceeds as follows. A starting vector $\mathbf{x}^0 \equiv \{x_1^0, \ldots x_M^0\}$ is drawn at random, and an initial "temperature" $kT$ is chosen such that it is comparable in value to the variation $\Delta U \equiv U_{max} - U_{min}$. Accordingly, a MC random walk will touch all regions of variable space with almost equal probability. If the temperature is now carefully lowered, the entire $\mathbf{x}$-space will still remain accessible at first, but regions with lower $U(\mathbf{x})$ will be visited more frequently than the higher ranges. Finally, for $kT \to 0$ the system point will come to rest in a minimum that very probably (albeit not with certainty) will be the global minimum.

Kirkpatrick and co-authors applied this technique to the minimization of electric leads in highly integrated electronic modules. Even at their very first attempt they achieved a considerable saving in computing time as compared to the proven optimization packages used until then.[KIRKPATRICK 83]

EXERCISE: Create (fake!) a table of "measured values with errors" according to

$$y_i = f(x_i; c_1, \ldots c_6) + \xi_i, \quad i = 1, 20 \tag{3.107}$$

with $\xi_i$ coming from a Gauss distribution with suitable variance, and with the function $f$ defined by

$$f(x; \mathbf{c}) \equiv c_1 e^{-c_2(x - c_3)^2} + c_4 e^{-c_5(x - c_6)^2} \tag{3.108}$$

($c_1 \ldots c_6$ being a set of arbitrary coefficients).

Using these data, try to reconstruct the parameters $c_1 \ldots c_6$ by fitting the theoretical function $f$ to the table points $(x_i, y_i)$. The cost function is

$$U(\mathbf{c}) \equiv \sum_i [y_i - f(x_i; \mathbf{c})]^2 \tag{3.109}$$

Choose an initial vector $\mathbf{c}^0$ and perform an MC random walk through $\mathbf{c}$-space, slowly lowering the temperature.

## 3.4.2   Genetic Algorithms

The evolution of biological systems is related to optimization in at least two respects. One, the adaptation of species to external conditions may be interpreted in terms of an optimization process. Two, the adaptation strategy itself has evolved over time, from the simple selective multiplication of prebiotic molecular systems to the sophisticated recipe of sexual reproduction used by eukaryotic organisms. Given the apparent success of the latter method, it is worthwhile to explore its performance in the setting of computational optimization.

For simplicity, consider some oscillatory function $f(x)$ of a single variable, having one global minimum within the range of definition, $x \epsilon [a, b]$. The solution of the minimization problem is then a number $x^*$ with $f(x^*) = min\{f(x), x \epsilon [a, b]\}$.

A genetic strategy to find $x^*$ proceeds as follows:

1. Start with a *population* of randomly chosen numbers (*individuals*), $\{x_i^0 \epsilon [a, b], i = 1, \ldots N\}$. The size $N$ of the population ($N = 100$, say) will be kept constant throughout the calculation. The *bit string* representing any of the members $x_i^0$ is understood as the "gene" of that individual which competes in "fitness" with all other $x_j^0$. In our simple example the fitness is bound to the value $f_i \equiv f(x_i^0)$: the lower $f_i$, the higher the fitness of $x_i^0$. It is always possible, and convenient, to assign the fitness such that it is positive definite.

   A relative fitness, or probability of reproduction, is defined as $p_i \equiv f_i / \sum_{i=1}^N f_i$. It has all the markings of a probability density, and accordingly we may also define a cumulative distribution function, $P(x_i) \equiv P_i \equiv \sum_{j=1}^{i} p_j$ (see equs. 3.15-3.16).

2. Next, draw $N$ individuals in accordance with their reproduction probability, allowing for repeated occurence of the same member. The proven recipe for this step is the well-known transformation (or inversion) method of Section 3.2.2: () draw a random number $\xi$ equidistributed in $[0,1]$; () put $P(x_i) = \xi$ and identify that $x_i$ for which this is true.

   Obviously, the new population $\{x'_i, \ i = 1, \ldots N\}$ will as a whole be fitter than the original one. However, thus far we have remained at the level of primitive selective reproduction without mutation or sexual crossover.

3. Pairs of individuals are now picked at random, and their genetic strings are submitted to crossover. In the simplest variant this is done as follows: () Draw a position $m$ within the bit strings; () swap the bits following $m$ between the two strings. The number of such pairings, the "crossover rate", should be around $0.6\,N$. The resulting set $\{x''_i, \ i = 1, \ldots N\}$ is called the *offspring* population.

4. Finally, *mutation* comes into play: within each string $x''_i$ every single bit is reversed with a probability $p_{mut} \approx 0.01$.

   The resulting population is regarded as the next generation, $\{x^1_i, \ i = 1, \ldots N\}$, and we are back at step 2.

A thorough textbook on the history, theory and practice of genetic algorithms is [GOLDBERG 89], and a fairly recent review is [TOMASSINI 95]. However, genetic algorithms are very much *en vogue*, and are rapidly improved, modified and applied in ever more fields. The only way to remain abreast of this development is a web search.

EXERCISE: Apply the simple genetic algorithm to find the minimum of the function $[2\,sin(10\,x-1)]^2+10\,(x-1)^2$ within the interval $[0,2]$.

# Part II

# Everything Flows

If it is true that mathematics is the language of physics, then differential equations surely are the verbs in it. It is therefore appropriate to devote part of this text to the numerical treatment of ordinary and partial differential equations.

We cannot fully understand today what an upheaval the discovery of the "fluxion", or differential, calculus must have been in its time. For us it is a matter of course to describe a certain model of growth by the equation

$$\dot{y}(t) = ay(t)$$

and to write down immediately the solution $y(t) \propto \exp(at)$, i.e. the notorious formula of exponential growth. Equally familiar is the concise Newtonian formulation of the mechanical law of motion,

$$\ddot{x}(t) = \frac{1}{m}K(t)$$

Only when we happen to come across an ancient text on ballistics, and find quite abstruse conceptions of the trajectories of cannon balls, we can sense how difficult the discussion of even such a simple physical problem as projectile motion must have been when the tools of differential calculus were not yet available.

Scientists were duly fascinated by the new methods. The French mathematicians and physicists of the eighteenth century brought "le calcul" to perfection and applied it to ever more problems. The sense of power they experienced found its expression in exaggerated announcements of an all-encompassing mechanical theory of all observable phenomena. No severe hindrance was seen in the fact that while for many phenomena one may well write down equations of motion, these may seldom be solved in explicit, "closed" form. "In principle" the solution was contained in the equations, everything else being a technical matter only.

At times the high esteem of infinitesimal calculus – or rather, the relatively poor image of algebra – would lead to remarkable mistakes. Thus the powerful opponent of Christian Doppler, the Viennese mathematician Petzval[4], derided the Doppler principle mostly for the reason that it was formulated as a simple algebraic relation and not as a differential or integral law.

Yet it is true: as every student of physics soon finds out, almost all relevant physical relations may be put in terms of differential equations. (This predominance of differential equations may in fact be due to our innate preference for linear-causal thinking; regrettably, this is not the place to discuss such matters.) And if we only decide to content ourselves with purely numerical solutions, we gain access to a whole world of phenomena by far transcending the class of simple cases analysable "in closed form".

The first step towards such a numerical solution is always a reformulation of the given differential equation in terms of a difference equation. A neologism describing

---

[4] JOSEF PETZVAL, 1807-91, co-founder of the "Chemico-Physical Society at Vienna" still in existence today. He became renowned for his numerical calculations on photographic multi-lens objectives, a project that makes him one of the forefathers of computational physics.

this step is "to difference" the respective equation. For instance, by replacing in

$$\frac{dx}{dt} = f(x)$$

the differential quotient by a difference quotient one obtains a linear equation, which in the most naive approximation reads

$$\frac{x_{n+1} - x_n}{\Delta t} \approx f(x_n)$$

Here $x_n \equiv x(t_n)$, and the time increment $\Delta t \equiv t_{n+1} - t_n$ is taken to be constant, i.e. independent of $n$. Obviously one may then, for given $x_n$ and $f(x_n)$, compute the next value $x_{n+1}$ according to

$$x_{n+1} \approx x_n + f(x_n)\,\Delta t$$

Iterative algorithms of this kind – albeit somewhat more refined and accurate – provide the basis for all classical and semi-classical simulation methods, as far as these presuppose deterministic equations of motion.

While the difference calculus suffices for the numerical treatment of ordinary differential equations, in the case of partial differential equations one has to invoke linear algebra as well. Since the solution function $u$ of such an equation depends on at least 2 variables, by discretizing those variables we obtain a table of functional values with 2 or more indices: $\{u_{i,j},\ i,j = 1,\ldots\}$. The given differential equation transforms into a set of difference equations which may be written as a matrix equation (see also Section 2.4).

# Chapter 4

# Ordinary Differential Equations

*Leonhard Euler provided the basic integration scheme*

An ordinary differential equation (ODE) in its most general form reads

$$L(x, y, y', y'', \ldots y^{(n)}) = 0 \tag{4.1}$$

where $y(x)$ is the solution function and $y' \equiv dy/dx$ etc. Most differential equations that are important in physics are of first or second order, which means that they contain no higher derivatives such as $y'''$ or the like. As a rule one may rewrite them in explicit form, $y' = f(x, y)$ or $y'' = g(x, y)$. Sometimes it is profitable to reformulate a given second-order DE as a system of two coupled first-order DEs. Thus, the equation of motion for the harmonic oscillator, $d^2x/dt^2 = -\omega_0^2 x$, may be transformed into the system of equations

$$\frac{dx}{dt} = v; \quad \frac{dv}{dt} = -\omega_0^2 x \tag{4.2}$$

Another way of writing this is

$$\frac{d\mathbf{y}}{dt} = \mathbf{L} \cdot \mathbf{y}, \quad \text{where } \mathbf{y} \equiv \begin{pmatrix} x \\ v \end{pmatrix} \text{ and } \mathbf{L} = \begin{pmatrix} 0 & 1 \\ -\omega_0^2 & 0 \end{pmatrix} \tag{4.3}$$

89

As we can see, $\mathbf{y}$ and $dy/dt$ occur only to first power: we are dealing with a *linear* differential equation.

Since the solution of a DE is determined only up to one or more constants, we need additional data in order to find the relevant solution. The number of such constants equals the number of formal integrations, i.e. the order of the DE. If the values of the required function and of its derivatives are all given at one single point $x_0$, we are confronted with an *initial value problem*. In contrast, if the set of necessary parameters is divided into several parts that are given at several points $x_0, x_1, \ldots$, we are dealing with a *boundary value problem*.

Typical initial value problems (IVP) are the various *equations of motion* to be found in all branches of physics. It is plausible that the conceptual basis of such equations is the idea that at some point in time the dynamical system can be known in all its details ("prepared"); the further evolution of the system is then given by the solution $y(t)$ of the equation of motion under the given initial condition.

As a standard example for boundary value problems (BVP) let us recall the equation governing the distribution of temperature along a thin rod. It reads $\lambda \, d^2T/dx^2 = 0$, and the two constants that define a unique solution are usually the temperature values at the ends of the rod, $T(x_0)$ and $T(x_1)$.

The distinction between IVP and BVP is quite superficial. It is often possible to reformulate an equation of motion as a BVP (as in ballistics), and a BVP may always be reduced to an IVP with initial values that are at first estimated and later corrected (see Sec. 4.3.1). However, the numerical techniques for treating the two classes of problems are very different.

# 4.1   Initial Value Problems of First Order

As mentioned before, initial value problems occur mainly in conjunction with equations of motion. We will therefore denote the independent variable by $t$ instead of $x$. The generic IVP of first order then reads

$$\frac{d\mathbf{y}}{dt} = \mathbf{f}(\mathbf{y}, t), \quad \text{with } \mathbf{y}(t = 0) = \mathbf{y}_0 \qquad (4.4)$$

To develop a numerical algorithm for solving this problem, let us apply the machinery of finite differences. First we discretize the $t$-axis, writing $\mathbf{y}_n \equiv \mathbf{y}(n\,\Delta t)$ and $\mathbf{f}_n \equiv \mathbf{f}(\mathbf{y}_n)$. The various formulae of Section 1.1 then provide us with several difference schemes – of varying quality – for determining $\mathbf{y}_1, \mathbf{y}_2$, etc.

## 4.1.1   Euler-Cauchy Algorithm

Recall the DNGF approximation to the first derivative of a tabulated function,

$$\left.\frac{d\mathbf{y}}{dt}\right|_{t_n} = \frac{\Delta \mathbf{y}_n}{\Delta t} + O[(\Delta t)] \qquad (4.5)$$

Inserting this in the given differential equation we obtain the difference equation

$$\frac{\Delta \mathbf{y}_n}{\Delta t} = f_n + O[(\Delta t)] \tag{4.6}$$

which immediately yields the *Euler-Cauchy* algorithm

$$\boxed{\mathbf{y}_{n+1} = \mathbf{y}_n + \mathbf{f}_n \Delta t + O[(\Delta t)^2]} \tag{4.7}$$

The obvious charm of this basic integration scheme is its algebraic and computational simplicity. However, we can see that it is accurate to first order only. An
even worse flaw is that for certain $\mathbf{f}(\mathbf{y})$ the EC method is not even stable, so that
small aberrations from the true solution tend to grow in the course of further steps.
We will demonstrate the phenomenon of instability of a difference scheme by way
of a simple example.

The *relaxation* or *decay* equation

$$\frac{dy(t)}{dt} = -\lambda y(t) \tag{4.8}$$

describes an exponential decrease or increase of the quantity $y(t)$, depending on
the sign of the parameter $\lambda$. The Euler-Cauchy formula for this DE reads

$$y_{n+1} = (1 - \lambda \Delta t) \, y_n \tag{4.9}$$

Of course, this formula will work better the smaller the time step $\Delta t$ we are using.
The error per time step – the "local error", which increases with $(\Delta t)^2$ – will
then be small. Indeed the numerical solution obtained with $\lambda \Delta t = 0.1$ is almost
indiscernible from the exact solution $y(t)/y_0 = exp(-\lambda t)$ (see Fig. 4.1). For
$\lambda \Delta t = 0.5$ the numerical result clearly deviates from the exponential. $\lambda \Delta t = 1.5$
and 2.0 result in sawtooth curves that differ quite far from the correct function,
but at least remain finite. For even larger values of $\lambda \Delta t$ the numerical solution –
and therefore the error – increases with each step.

## 4.1.2    Stability and Accuracy of Difference Schemes

What happened? The following stability analysis permits us to determine, for a
given DE and a specific numerical algorithm, the range of stability, i.e. the largest
feasible $\Delta t$. As a rule the rationale for choosing a small time step is to achieve a
high accuracy per step (i.e. a small local error.) But there are cases where an ever
so small $\Delta t$ leads, in the course of many steps, to a "secular", systematic increase
of initially small deviations. Stability analysis allows us to identify such cases by
returning the verdict "zero stability range."

We denote by $\mathbf{y}(t)$ the – as a rule unknown – exact solution of the given DE,
and by $\mathbf{e}(t)$ an error that may have accumulated in our calculation. In other words,

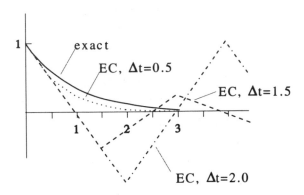

Figure 4.1: Solutions to the equation $dy/dt = -\lambda y$, with $\lambda = 1$ and $y_0 = 1$

our algorithm has produced the approximate solution $\mathbf{y}_n + \mathbf{e}_n$ at time $t_n$. What, then, is the approximate solution at time $t_{n+1}$? For the EC method we have

$$\mathbf{y}_{n+1} + \mathbf{e}_{n+1} = \mathbf{y}_n + \mathbf{e}_n + \mathbf{f}(\mathbf{y}_n + \mathbf{e}_n)\Delta t \qquad (4.10)$$

The EC formula is the most basic member of a class of so-called *single step algorithms*, which produce the solution at time $t_{n+1}$ by application of some transformation $T$ to the value of the solution at time $t_n$:

$$\mathbf{y}_{n+1} + \mathbf{e}_{n+1} = T(\mathbf{y}_n + \mathbf{e}_n) \qquad (4.11)$$

Assuming that the deviation $\mathbf{e}_n$ is small and the transformation $T$ is well-behaved, we may expand $T(\mathbf{y}_n + \mathbf{e}_n)$ around the correct solution $\mathbf{y}_n$:

$$T(\mathbf{y}_n + \mathbf{e}_n) \approx T(\mathbf{y}_n) + \left.\frac{dT(\mathbf{y})}{d\mathbf{y}}\right|_{\mathbf{y}_n} \cdot \mathbf{e}_n \qquad (4.12)$$

Since $T(\mathbf{y}_n) = \mathbf{y}_{n+1}$, we have from 4.11

$$\mathbf{e}_{n+1} \approx \left.\frac{dT(\mathbf{y})}{d\mathbf{y}}\right|_{\mathbf{y}_n} \cdot \mathbf{e}_n \equiv \mathbf{G} \cdot \mathbf{e}_n \qquad (4.13)$$

The matrix $\mathbf{G}$ is called *amplification matrix*. Obviously the repeated multiplication of some initial error $\mathbf{e}_0$ (which may simply be caused by the finite number of digits in a computer word) may lead to diverging error terms. Such divergences will be absent only if all eigenvalues of $\mathbf{G}$ are situated within the unit circle:

$$|g_i| \leq 1, \quad \text{for all } i \qquad (4.14)$$

Let us apply this insight to the above example of the relaxation equation. In the Euler-Cauchy method 4.9 the transformation $T$ is simply a multiplication by the factor $(1 - \lambda \Delta t)$:

$$T(y_n) \equiv (1 - \lambda \Delta t)\, y_n \qquad (4.15)$$

The amplification "matrix" $\mathbf{G}$ then degenerates to the scalar quantity $(1- \lambda \Delta t)$, and the range of stability is defined by the requirement that

$$|1 - \lambda \Delta t| \leq 1 \tag{4.16}$$

For $\lambda = 1$ this condition is met whenever $\Delta t \leq 2$. Indeed, it was just the limiting value $\Delta t = 2$ which produced the marginally stable sawtooth curve in Figure 4.1.

EXAMPLE: As a less trivial example for the application of stability analysis we will consider the harmonic oscillator. Applying the Euler-Cauchy scheme to 4.3 we find

$$\mathbf{y}_{n+1} = [\mathbf{I} + \mathbf{L} \Delta t] \cdot \mathbf{y}_n \equiv T(\mathbf{y}_n) \tag{4.17}$$

The amplification matrix is

$$\mathbf{G} \equiv \left. \frac{dT(\mathbf{y})}{d\mathbf{y}} \right|_{\mathbf{y}_n} = \mathbf{I} + \mathbf{L} \Delta t \tag{4.18}$$

The eigenvalues of $\mathbf{G}$ are $g_{1,2} = 1 \pm i\omega_0 \Delta t$, so that

$$|g_{1,2}| = \sqrt{1 + (\omega_0 \, \Delta t)^2} \tag{4.19}$$

Regardless how small we choose $\Delta t$, we have always $|g_{1,2}| > 1$. We conclude that the EC method applied to the harmonic oscillator is never stable.

In the following descriptions of several important algorithms the range of stability will in each instance be given for the two standard equations – relaxation and harmonic oscillator. A more in-depth discussion of the stability of various methods for initial value problems may be found in [GEAR 71]. For completeness, here follow a few concepts that are helpful in discussing the stability and accuracy of iterative methods:

Let $L(y) = 0$ be the given DE, with the exact solution $y(t)$. (Example: $L(y) \equiv \dot{y} + \lambda y = 0$; relaxation equation.) Also, let $F(y) = 0$ be a truncated difference scheme pertaining to the given DE, with its own *exact* solution $y_n$. (Example: $y_n$ as computed by repeated application of 4.9.)

**Cumulative truncation error:** This is the difference, at time $t_n$, between the solution of the DE and that of the difference equation:

$$e_n \equiv y(t_n) - y_n \tag{4.20}$$

**Convergence:** A difference scheme is convergent if its solution approaches for decreasing time steps the solution of the DE:

$$\lim_{\Delta t \to 0} y_n = y(t_n) \quad \text{or} \quad \lim_{\Delta t \to 0} e_n = 0 \tag{4.21}$$

**Local truncation error:** Inserting the exact solution of the DE in the difference scheme one usually obtains a finite value, called the local truncation error:

$$F_n \equiv F[y(t_n)] \tag{4.22}$$

**Consistency:** The algorithm $F(y) = 0$ is consistent if

$$\lim_{\Delta t \to 0} F_n = 0 \tag{4.23}$$

**Roundoff error:** Due to the finite accuracy of the representation of numbers (for example, but not exclusively, in the computer) the practical application of the difference scheme yields, instead of $y_n$, a somewhat different value $\bar{y}_n$. The discrepancy is called roundoff error:

$$r_n = \bar{y}_n - y_n \tag{4.24}$$

**Stability:** The ubiquitous roundoff errors may "excite" a solution of the difference equation that is not contained in the original DE. If in the course of many iterations this undesired solution grows without bounds, the method is unstable.

## 4.1.3  Explicit Methods

The Euler-Cauchy formula is the most simple example of an *explicit* integration scheme. These are procedures that use an explicit expression for $y_{n+1}$ in terms of $y$ and $f$ as given from preceding time steps. (If only $y_n$ and $f_n$ occur, as in the EC method, we are dealing with an explicit *single step* scheme.)

The EC formula was derived using that difference quotient which in Section 1.2 was called DNGF approximation. We may obtain another explicit scheme by introducing the DST approximation:

$$\left.\frac{dy}{dt}\right|_{t_n} \approx \frac{1}{\Delta t}\mu\delta y_n = \frac{1}{2\Delta t}[y_{n+1} - y_{n-1}] \tag{4.25}$$

The DE $dy/dt = f(t)$ is thus transformed into a sequence of difference equations,

$$
\begin{aligned}
y_{n+1} &= y_{n-1} + f_n\, 2\Delta t + O[(\Delta t)^3] &(4.26)\\
y_{n+2} &= y_n + f_{n+1}\, 2\Delta t + O[(\Delta t)^3] &(4.27)\\
&\text{etc.}
\end{aligned}
$$

Each line is an explicit formula of first order that couples the values of $y$ at time steps $t_{n+1}$ and $t_{n-1}$, omitting the quantity $y_n$. However, $f_n \equiv f(y_n)$ is needed and has to be evaluated in the preceding step. This two-step procedure is pictorially called *leapfrog* technique.

Note that on the right hand side of 4.26 there appear *two* time steps. The stability analysis of such *multistep techniques* is a straightforward generalization of the method explained before. Let us write the general form of an explicit multistep scheme as

$$\mathbf{y}_{n+1} = \sum_{j=0}^{k} \left[ a_j \mathbf{y}_{n-j} + b_j \Delta t\, \mathbf{f}_{n-j} \right] \tag{4.28}$$

Applying the same formula to a slightly deviating solution $\mathbf{y}_{n-j} + \mathbf{e}_{n-j}$ and computing the difference, we have in linear approximation

$$\mathbf{e}_{n+1} \approx \sum_{j=0}^{k} \left[ a_j \mathbf{I} + b_j \Delta t \left. \frac{d\mathbf{f}}{d\mathbf{y}} \right|_{\mathbf{y}_n} \right] \cdot \mathbf{e}_{n-j} \equiv \sum_{j=0}^{k} \mathbf{A}_j \cdot \mathbf{e}_{n-j} \tag{4.29}$$

Defining the new error vectors

$$\boldsymbol{\eta}_n \equiv \begin{pmatrix} \mathbf{e}_n \\ \mathbf{e}_{n-1} \\ \vdots \\ \mathbf{e}_{n-k} \end{pmatrix} \tag{4.30}$$

and the quadratic matrix

$$\mathbf{G} \equiv \begin{pmatrix} \mathbf{A}_0 & \mathbf{A}_1 & \cdots & \mathbf{A}_k \\ \mathbf{I} & 0 & \cdots & 0 \\ 0 & \ddots & & 0 \\ 0 & \cdots & \mathbf{I} & 0 \end{pmatrix} \tag{4.31}$$

we may write the law of error propagation in the same form as 4.13,

$$\boldsymbol{\eta}_{n+1} = \mathbf{G} \cdot \boldsymbol{\eta}_n \tag{4.32}$$

Again, the stability criterion reads

$$|g_i| \le 1, \quad \text{for all } i \tag{4.33}$$

EXAMPLE 1: Applying the leapfrog scheme to the relaxation equation one obtains the scalar formula

$$y_{n+1} = y_{n-1} - 2\Delta t \lambda y_n + O[(\Delta t)^3] \tag{4.34}$$

The error propagation obeys

$$e_{n+1} \approx -2\Delta t \lambda e_n + e_{n-1} \tag{4.35}$$

so that $A_0 = -2\Delta t\lambda$, and $A_1 = 1$. The matrix $\mathbf{G}$ is therefore given by

$$\mathbf{G} = \begin{pmatrix} -2\Delta t\lambda & 1 \\ 1 & 0 \end{pmatrix} \tag{4.36}$$

with eigenvalues

$$g_{1,2} = -\lambda\Delta t \pm \sqrt{(\lambda\Delta t)^2 + 1} \tag{4.37}$$

Since in the relaxation equation the quantity $\lambda\Delta t$ is real, we have $|g_2| > 1$ under all circumstances. The leapfrog scheme is therefore unsuitable for treating decay or growth problems.

EXAMPLE 2: If we apply the leapfrog method to the harmonic oscillator, we obtain (using the definitions of equ. 4.3)

$$\mathbf{y}_{n+1} = 2\Delta t\, \mathbf{L} \cdot \mathbf{y}_n + \mathbf{y}_{n-1} \tag{4.38}$$

and consequently

$$\mathbf{e}_{n+1} \approx 2\Delta t\, \mathbf{L} \cdot \mathbf{e}_n + \mathbf{e}_{n-1} \tag{4.39}$$

The amplification matrix is therefore, with $\alpha \equiv 2\Delta t$,

$$\mathbf{G} = \begin{pmatrix} \alpha\mathbf{L} & \mathbf{I} \\ \mathbf{I} & \mathbf{0} \end{pmatrix} = \begin{pmatrix} 0 & \alpha & 1 & 0 \\ -\alpha\omega_0^2 & 0 & 0 & 1 \\ 1 & 0 & 0 & 0 \\ 0 & 1 & 0 & 0 \end{pmatrix} \tag{4.40}$$

For the eigenvalues of $\mathbf{G}$ we find

$$g = \pm \left[ (1 - \frac{\alpha^2\omega_0^2}{2}) \pm i\alpha\omega_0 \sqrt{1 - \frac{\alpha^2\omega_0^2}{4}} \right]^{1/2} \tag{4.41}$$

so that

$$|g| = 1. \tag{4.42}$$

Thus the algorithm, when applied to the harmonic oscillator, is *marginally stable*, regardless of the specific values of $\Delta t$ and $\omega_0^2$.

## 4.1.4   Implicit Methods

The most fundamental *implicit* scheme is obtained by approximating the time derivative by the DNGB (instead of the DNGF) formula:

$$\frac{dy}{dt}\bigg|_{n+1} = \frac{\nabla \mathbf{y}_{n+1}}{\Delta t} + O[\Delta t] \tag{4.43}$$

Inserting this in $dy/dt = f[\mathbf{y}(t)]$ we find

$$\mathbf{y}_{n+1} = \mathbf{y}_n + \mathbf{f}_{n+1}\Delta t + O[(\Delta t)^2] \tag{4.44}$$

This formula is of first order accuracy only, no more than the explicit Euler-Cauchy scheme, but as a rule it is much more stable. The problem is that the quantity $\mathbf{f}_{n+1}$ is not known at the time it would be needed – namely at time $t_n$. Only if $\mathbf{f}(\mathbf{y})$ is a *linear* function of its argument $\mathbf{y}$ are we in a position to translate 4.44 into a feasible integration algorithm. Writing $\mathbf{f}_{n+1} = \mathbf{L} \cdot \mathbf{y}_{n+1}$, we then have

$$\mathbf{y}_{n+1} = [\mathbf{I} - \mathbf{L}\Delta t]^{-1} \cdot \mathbf{y}_n + O[(\Delta t)^2] \tag{4.45}$$

The higher stability of this method as compared to the Euler formula may be demonstrated by way of our standard problems. The evolution of errors obeys

$$\mathbf{e}_{n+1} = [\mathbf{I} - \mathbf{L}\Delta t]^{-1} \cdot \mathbf{e}_n \equiv \mathbf{G} \cdot \mathbf{e}_n \tag{4.46}$$

For the relaxation equation $\mathbf{G} = G = 1/(1 + \lambda\Delta t)$, and obviously $|g| < 1$ for any $\lambda > 0$. (On first sight the case $\lambda < 0$ seems to be dangerous; but then we are dealing with a *growth* equation, and the *relative* error $e/y$ will still remain bounded.) In the case of the harmonic oscillator we have

$$\mathbf{G} \equiv [\mathbf{I} - \mathbf{L}\Delta t]^{-1} = \frac{1}{1 + (\omega_0\Delta t)^2} \begin{pmatrix} 1 & \Delta t \\ -\omega_0^2\Delta t & 1 \end{pmatrix} \tag{4.47}$$

with eigenvalues

$$g_{1,2} = \frac{1}{1 + (\omega_0\Delta t)^2}[1 \pm i\omega_0\Delta t] \tag{4.48}$$

so that

$$|g|^2 = \frac{1}{1 + (\omega_0\Delta t)^2} \tag{4.49}$$

which is smaller than 1 for any $\Delta t$.

An implicit scheme of *second order* may be obtained in the following manner. We truncate the DNGF approximation 1.30 after the second term and write it down for $u = 0$ (i.e. $t = t_n$) and for $u = 1$ (meaning $t_{n+1}$), respectively:

$$\mathbf{f}_n \equiv \dot{\mathbf{y}}(t_n) = \frac{1}{\Delta t}[\Delta \mathbf{y}_n - \frac{1}{2}\Delta^2 \mathbf{y}_n] + O[(\Delta t)^2] \tag{4.50}$$

$$\mathbf{f}_{n+1} \equiv \dot{\mathbf{y}}(t_{n+1}) = \frac{1}{\Delta t}[\Delta \mathbf{y}_n + \frac{1}{2}\Delta^2 \mathbf{y}_n] + O[(\Delta t)^2] \tag{4.51}$$

Adding the two lines yields

$$\mathbf{y}_{n+1} = \mathbf{y}_n + \frac{\Delta t}{2}[\mathbf{f}_n + \mathbf{f}_{n+1}] + O[(\Delta t)^3] \tag{4.52}$$

Again, this implicit formula can be of any practical use only if $\mathbf{f}$ is linear in $\mathbf{y}$. With $\mathbf{f}_n = \mathbf{L} \cdot \mathbf{y}_n$ etc. we obtain from 4.52

$$\mathbf{y}_{n+1} = [\mathbf{I} - \mathbf{L}\frac{\Delta t}{2}]^{-1} \cdot [\mathbf{I} + \mathbf{L}\frac{\Delta t}{2}] \cdot \mathbf{y}_n + O[(\Delta t)^3] \tag{4.53}$$

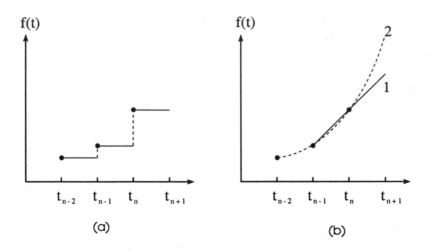

Figure 4.2: PC method: a) EC ansatz: step function for $f(t)$; b) general predictor-corrector schemes: 1 ... linear NGB extrapolation; 2 ... parabolic NGB extrapolation

Stability is guaranteed for the decay equation if

$$|g| \equiv \left| \frac{1 - \lambda \, \Delta t/2}{1 + \lambda \, \Delta t/2} \right| \leq 1 \tag{4.54}$$

which is always true for $\lambda > 0$. For the harmonic oscillator

$$g_{1,2} \equiv \frac{1 \pm i \, \omega_0 \Delta t/2}{1 + (\omega_0 \Delta t)^2/4} \tag{4.55}$$

with $|g| \leq 1$ for all $\Delta t$.

### 4.1.5   Predictor-Corrector Method

The explicit and implicit schemes explained in the preceding sections are of first and second order only. In many applications this is not good enough. The following predictor-corrector schemes provide a systematic extension towards higher orders of accuracy. In this context the predictor is an *explicit* formula, while the corrector may be seen as a kind of *implicit* prescription.

To understand the way in which predictors of arbitrary order are constructed we once more consider the simple EC formula. Equation 4.7 is based on the assumption that the kernel $f(t)$ maintains the value $f_n$ for the entire period $[t_n, t_{n+1}]$ (see Fig. 4.2a). It is evident that for a systematic improvement we simply have to replace this step function by an extrapolation polynomial of order $1, 2, \ldots$ using the values of $f_n, f_{n-1}, f_{n-2} \ldots$ (Fig. 4.2b). The general NGB polynomial

$$f(t_n + \tau) = f_n + \frac{u}{1!} \nabla f_n + \frac{u(u+1)}{2!} \nabla^2 f_n + \ldots \tag{4.56}$$

(with $u \equiv \tau/\Delta t$) is thus extended into the time interval $[t_n, t_{n+1}]$. This renders the right-hand side of the DE $dy/dt = f(t)$ formally integrable, and we obtain according to

$$y_{n+1}^P = y_n + \Delta t \int_0^1 du\, f(t_n + u\Delta t) \tag{4.57}$$

the general *Adams-Bashforth predictor*

$$y_{n+1}^P = y_n + \Delta t \left[ f_n + \frac{1}{2}\nabla f_n + \frac{5}{12}\nabla^2 f_n + \frac{3}{8}\nabla^3 f_n + \right.$$
$$\left. + \frac{251}{720}\nabla^4 f_n + \frac{95}{288}\nabla^5 f_n + \ldots \right] \tag{4.58}$$

Depending on how far we go with this series we obtain the various predictor formulae listed in Table 4.1. The predictor of first order is, of course, just the Euler-Cauchy formula; the second order predictor is often called *open trapezoidal rule*.

As soon as the predictor $y_{n+1}^P$ is available we may perform the *evaluation step* to determine the quantity

$$f_{n+1}^P \equiv f[y_{n+1}^P] \tag{4.59}$$

which will usually deviate somewhat from the value of the extrapolation polynomial 4.56 at time $t_{n+1}$. Now inserting $f_{n+1}^P$ in a *backward* interpolation formula around $t_{n+1}$, we can expect to achieve a better approximation than by the original extrapolation – albeit within the same order of accuracy. Once more we may integrate analytically,

$$y_{n+1} = y_n + \Delta t \int_{-1}^0 du\, f(t_{n+1} + u\Delta t) \tag{4.60}$$

to obtain the general *Adams-Moulton corrector*

$$y_{n+1} = y_n + \Delta t \left[ f_{n+1} - \frac{1}{2}\nabla f_{n+1} - \frac{1}{12}\nabla^2 f_{n+1} - \frac{1}{24}\nabla^3 f_{n+1} \right.$$
$$\left. - \frac{19}{720}\nabla^4 f_{n+1} - \frac{3}{160}\nabla^5 f_{n+1} - \ldots \right] \tag{4.61}$$

(where $\nabla f_{n+1} \equiv f_{n+1}^P - f_n$ etc.). The first few correctors of this kind are assembled in Table 4.2.

A final *evaluation step* $f_{n+1} \equiv f(y_{n+1})$ yields the definitive value of $f_{n+1}$ to be used in the calculation of the next predictor. One might be tempted to insert the corrected value of $f_{n+1}$ once more in the corrector formula. The gain in accuracy, however, is not sufficient to justify the additional expense in computing time. Thus the PC method should always be applied according to the pattern PECE, i.e. "prediction-evaluation-correction-evaluation." An iterated procedure like P(EC)$^2$E is not worth the effort.

---

**Predictors for first order differential equations:**

$$y_{n+1}^P = y_n \;+\; \Delta t\, f_n + O[(\Delta t)^2] \tag{4.62}$$

$$\ldots \;+\; \frac{\Delta t}{2}[3f_n - f_{n-1}] + O[(\Delta t)^3] \tag{4.63}$$

$$\ldots \;+\; \frac{\Delta t}{12}[23f_n - 16f_{n-1} + 5f_{n-2}] + O[(\Delta t)^4] \tag{4.64}$$

$$\ldots \;+\; \frac{\Delta t}{24}[55f_n - 59f_{n-1} + 37f_{n-2} - 9f_{n-3}] + O[(\Delta t)^5] \tag{4.65}$$

$$\vdots$$

---

Table 4.1: Adams-Bashforth predictors

---

**Correctors for first order differential equations:**

$$y_{n+1} = y_n \;+\; \Delta t\, f_{n+1}^P + O[(\Delta t)^2] \tag{4.66}$$

$$\ldots \;+\; \frac{\Delta t}{2}[f_{n+1}^P + f_n] + O[(\Delta t)^3] \tag{4.67}$$

$$\ldots \;+\; \frac{\Delta t}{12}[5f_{n+1}^P + 8f_n - f_{n-1}] + O[(\Delta t)^4] \tag{4.68}$$

$$\ldots \;+\; \frac{\Delta t}{24}[9f_{n+1}^P + 19f_n - 5f_{n-1} + f_{n-2}] + O[(\Delta t)^5] \tag{4.69}$$

$$\vdots$$

---

Table 4.2: Adams-Moulton correctors

The PC methods may be thought of as a combination of explicit and implicit formulae. Accordingly the stability range is also intermediate between the narrow limits of the explicit and the much wider ones of the implicit schemes. Applying, for example, the Adams-Bashforth predictor of second order to the relaxation equation one finds for the eigenvalues of the amplification matrix $\mathbf{G}$ the characteristic equation

$$g^2 - g(1 - \frac{3}{2}\alpha) - \frac{\alpha}{2} = 0 \tag{4.70}$$

where $\alpha \equiv \lambda \Delta t$. For positive $\lambda$ we have $|g| \leq 1$, as long as $\Delta t \leq 1/\lambda$. The Adams-Moulton corrector on the other hand yields the error equation

$$e_{n+1} = \frac{1 - \alpha/2}{1 + \alpha/2} e_n \equiv g e_n \tag{4.71}$$

with $|g| < 1$ for all $\alpha > 0$, i.e. for any $\Delta t$ at all. The limit of stability for the combined method should therefore be situated somewhere between $\Delta t = 1/\lambda$ and $\Delta t = \infty$. This is indeed borne out by testing the method on the relaxation equation. Inserting the predictor formula of order 2,

$$y_{n+1}^P = y_n - \frac{\alpha}{2}(3y_n - y_{n-1}) \tag{4.72}$$

in the corrector formula

$$y_{n+1} = y_n - \frac{\alpha}{2}(y_{n+1}^P + y_n) \tag{4.73}$$

one finds

$$y_{n+1} = y_n(1 - \alpha + \frac{3}{4}\alpha^2) - \frac{\alpha^2}{4} y_{n-1} \tag{4.74}$$

with an identical error equation. The amplification factor thus obeys the equation

$$g^2 - g(1 - \alpha + \frac{3}{4}\alpha^2) + \frac{\alpha^2}{4} = 0 \tag{4.75}$$

For positive $\alpha \leq 2$ the solutions to this equation are situated within the unit circle; the condition for the time step is therefore $\Delta t \leq 2/\lambda$. The stability range of the combined method is thus twice as large as that of the bare predictor ($\Delta t \leq 1/\lambda$).

## 4.1.6   Runge-Kutta Method

To understand the idea of the RK technique we once more return to the simple EC formula 4.7. It rests on the assumption that $f(t)$ retains the value $f_n$ during the whole of the time interval $[t_n, t_{n+1}]$ (see Figure 4.3a). A more refined approach would be to calculate first a predictor for $y$ at half-time $t_{n+1/2}$, evaluate $f_{n+1/2} = f(y_{n+1/2})$ and then compute a kind of corrector at $t_{n+1}$ (Fig. 4.3b):

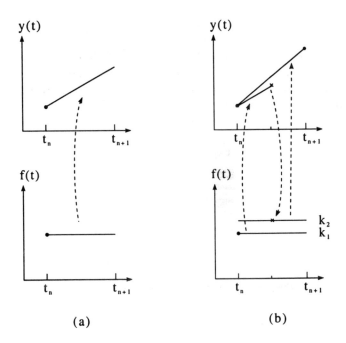

Figure 4.3: Runge-Kutta method. a) EC formula (= RK of first order); b) RK of second order

---

**Runge-Kutta of order 4 for first-order ODE:**

$$k_1 = \Delta t\, f(y_n)$$

$$k_2 = \Delta t\, f(y_n + \frac{1}{2}k_1)$$

$$k_3 = \Delta t\, f(y_n + \frac{1}{2}k_2)$$

$$k_4 = \Delta t\, f(y_n + k_3)$$

$$y_{n+1} = y_n + \frac{1}{6}[k_1 + 2k_2 + 2k_3 + k_4] + O[(\Delta t)^5] \qquad (4.77)$$

---

Table 4.3: Runge-Kutta of order 4

**Runge-Kutta of order 2:**

$$k_1 = \Delta t\, f(y_n)$$

$$k_2 = \Delta t\, f(y_n + \frac{1}{2}k_1)$$

$$y_{n+1} = y_n + k_2 + O[(\Delta t)^3] \qquad (4.76)$$

This algorithm is called *Runge-Kutta method of second order*, or *half-step method*. Yet another name for the same algorithm is *Euler-Richardson method*. It is related to the predictor-corrector technique of second order – with the difference that the quantity $f_{n-1}$ is not needed. Equation 4.76 is therefore a single step method and may accordingly be applied even at the first time step $t_0 \rightarrow t_1$; such an algorithm is called *self-starting*.

A much more powerful method that has found wide application is the RK algorithm of order 4, as described in Table 4.3.

The most important advantage of the RK method as compared to the PC algorithms is that at time $t_n$ no preceding values of $f_{n-1}, f_{n-1}, \ldots$ need be known. This is a valuable property not only for starting a calculation from $t_0$, but also for varying the time step in the course of the computation. If, for instance, the variation of $f[y(t)]$ goes up, $\Delta t$ may be decreased to keep numerical errors small. Also, the local truncation error may be estimated most easily by computing $y_{n+1}$ first with $\Delta t$ and then once more in two steps of length $\Delta t/2$.

One flaw of the RK method is the necessity of repeatedly evaluating $f(y)$ in one time step. Particularly in $N$-body simulations (molecular dynamics calculations) the evaluation step is very costly, and the RK method has never become popular with simulators.

Stability analysis for the RK algorithm proceeds along similar lines as for the PC methods. The half-step technique applied to the decay equation leads to an

---

**Extrapolation method:**

1. From a given (rather large) interval $\Delta t \equiv t_1 - t_0$ form successively smaller steps $h \equiv \Delta t/n$, with $n = 2, 4, 6, 8, 12, \ldots$, (in general, $n_j = 2n_{j-2}$.)

2. With each of these divided steps $h$ compute the table values

$$
\begin{aligned}
z_0 &= y_0 \\
z_1 &= z_0 + hf(z_0) \\
z_{m+1} &= z_{m-1} + 2hf(z_m); \quad m = 1, 2 \ldots n - 1 \quad (leapfrog!)
\end{aligned}
$$

and finally

$$y_1 = \frac{1}{2}[z_n + z_{n-1} + hf(z_n)]. \tag{4.79}$$

3. In this way a sequence of estimated end values $y_1$ are created that depend on the divided step width $h$: $y_1 = y_1(h)$. This sequence is now extrapolated towards vanishing step width, $h \to 0$. The best way to do this is *rational* extrapolation, meaning that one fits the given pairs $\{h, y_1(h)\}$ by a rational function

$$R(h) = \frac{P(h)}{Q(h)} \tag{4.80}$$

where $P$ and $Q$ are polynomials.

---

Figure 4.4: Extrapolation method by Bulirsch and Stoer

error propagation following

$$e_{n+1} = (1 - \alpha + \frac{\alpha^2}{2})e_n \equiv ge_n \tag{4.78}$$

with $\alpha \equiv \lambda\Delta t$. For positive $\lambda$ this implies $|g| \leq 1$ whenever $\Delta t \leq 2/\lambda$.

## 4.1.7   Extrapolation Method

When discussing the Runge-Kutta method we have already mentioned the possibility of estimating the local truncation error by subdividing the given time step $\Delta t$. The authors Richardson, Bulirsch, and Stoer [STOER 89, GEAR 71] have extended this idea and have forged it to a method which to a large extent eliminates that error. The principle of their method is sketched in Figure 4.4.

A thorough description of this extremely accurate and stable, but also rather expensive technique may be found in [STOER 89] and in [PRESS 86].

EXERCISE: Test various algorithms by applying them to an analytically solvable problem, as the harmonic oscillator or the 2-body Kepler problem. Include in your code tests that do not rely on the existence of an analytical solution (energy conservation or such.) Finally, apply the code to more complex problems such as the anharmonic oscillator or the many-body Kepler problem.

## 4.2   Initial Value Problems of Second Order

The fundamental equation of motion in classical point mechanics reads, in cartesian coordinates,

$$\frac{d^2\mathbf{r}}{dt^2} = \frac{1}{m}\mathbf{K}[\mathbf{r}(t)] \qquad (4.81)$$

Similar equations hold for the rotatory motion of rigid bodies or of flexible chains. And in almost all branches of physics we are faced with some paraphrase of the harmonic oscillator or the more general anharmonic oscillator

$$\frac{d^2y}{dt^2} = -\omega_0^2 y - \beta y^3 - \ldots \equiv b(y) \qquad (4.82)$$

Since in many instances the *acceleration b* may depend also on the *velocity dy/dt* – as in the presence of frictional or electromagnetic forces – we will write the second-order equation of motion in the general form

$$\frac{d^2y}{dt^2} = b[y, dy/dt] \qquad (4.83)$$

It was mentioned before that a second-order DE may always be rewritten as a system of two coupled equations of first order, so that the algorithms of the preceding section are applicable. However, there are several very efficient techniques that have been specially designed for the direct numerical integration of second-order differential equations.

### 4.2.1   Verlet Method

Loup Verlet introduced this technique in 1967 in the context of his pioneering molecular dynamics simulations on Argon [VERLET 67]. A different formulation of the same algorithm was introduced earlier by G. H. Vineyard [VINEYARD 62]. Although it contains terms up to $(\Delta t)^2$ only, the algorithm is of third order accuracy, which is sufficient for this type of simulations. A more accurate technique which shares the first three terms with Verlet's had been used as early as 1905 by the Norwegian mathematician C. Størmer to trace the capricious paths of charged elementary particles that are trapped in the magnetic field of the earth. Størmer performed these computations on the aurora problem together with several of his students, without any modern computing aids, in about 5000 hours of work – a true founding father of computational physics [STOERMER 07, STOERMER 21].

To derive the Verlet algorithm one simply replaces the second differential quotient by the Stirling approximation (see equ. 1.48):

$$\frac{d^2y}{dt^2}\bigg|_n = \frac{\delta^2 y_n}{(\Delta t)^2} + O[(\Delta t)^2] \tag{4.84}$$

This leads immediately to

$$y_{n+1} = 2y_n - y_{n-1} + b_n(\Delta t)^2 + O[(\Delta t)^4] \tag{4.85}$$

Note that the velocity $v \equiv \dot{y}$ does not appear explicitly. The *a posteriori* estimate for the velocity $v_n$,

$$v_n = \frac{1}{2\Delta t}[y_{n+1} - y_{n-1}] + O[(\Delta t)^2] \tag{4.86}$$

is quite inaccurate and may be used for crude checks only. Also, the Verlet algorithm is not self-starting. In addition to the initial value $y_0$ one needs $y_{-1}$ to tackle the first time step. In a typical initial value problem the quantities $y_0$ and $\dot{y}_0$ are given instead. By *estimating* some suitable $y_{-1}$ in order to start a Verlet calculation one solves not the given IVP but a very similar one. Still, the method has become very popular in statistical-mechanical simulation. It must be remembered that the aim of such simulations is not to find the *exact* solution to an accurately defined initial value problem, but to simulate the "typical" dynamics of an $N$-body system, for *approximately* given initial conditions.

If the Verlet method is to be applied to a problem with *exact* initial values, the first time step must be bridged by a self-starting technique, such as Runge-Kutta (see below.)

Stability analysis proceeds in a similar way as for the methods of Section 4.1. For our standard problem we will use the harmonic oscillator in its more common formulation as a DE of second order. The Verlet algorithm then reads

$$y_{n+1} = 2y_n - y_{n-1} - \omega_0^2 y_n(\Delta t)^2 \tag{4.87}$$

whence it follows that

$$e_{n+1} = (2 - \alpha^2)e_n - e_{n-1} \tag{4.88}$$

with $\alpha \equiv \omega_0 \Delta t$. The eigenvalue equation

$$|\mathbf{G} - g\mathbf{I}| = \begin{vmatrix} (2 - \alpha^2 - g) & -1 \\ 1 & -g \end{vmatrix} = 0 \tag{4.89}$$

reads

$$g^2 - (2 - \alpha^2)g + 1 = 0 \tag{4.90}$$

---

**Verlet leapfrog:**

$$v_{n+1/2} = v_{n-1/2} + b_n \Delta t \tag{4.92}$$

$$v_n = \frac{1}{2}(v_{n+1/2} + v_{n-1/2}) \quad \text{(if desired)} \tag{4.93}$$

$$y_{n+1} = y_n + v_{n+1/2}\Delta t + O[(\Delta t)^4] \tag{4.94}$$

---

Figure 4.5: Leapfrog version of the Verlet method

---

**Velocity Verlet:**

$$y_{n+1} = y_n + v_n\Delta t + b_n\frac{(\Delta t)^2}{2} + O[(\Delta t)^4] \tag{4.95}$$

$$v_{n+1/2} = v_n + b_n\frac{\Delta t}{2} \tag{4.96}$$

$$Evaluation\ step\ y_{n+1} \rightarrow b_{n+1} \tag{4.97}$$

$$v_{n+1} = v_{n+1/2} + b_{n+1}\frac{\Delta t}{2} \tag{4.98}$$

---

Figure 4.6: Swope's formulation of the Verlet algorithm

Its root

$$g = (1 - \frac{\alpha^2}{2}) \pm \sqrt{\frac{\alpha^4}{4} - \alpha^2} \tag{4.91}$$

is imaginary for $\alpha < 2$, with $|g|^2 = 1$. In the case $\alpha \geq 2$ – which for reasons of accuracy is excluded anyway – the Verlet algorithm would be unstable.

Incidentally, there are two further formulations of the Verlet method, which are known as the "leapfrog" – this is the one given by Vineyard – and "velocity Verlet" algorithms, respectively. We have already encountered a leapfrog method for treating differential equations of first order (see Sec. 4.1.3). Figure 4.5 shows the leapfrog scheme appropriate to second order DEs. It is important to note that in this formulation of the procedure the velocity – or rather, a crude estimate of $v$ – is available already at time $t_n$ (see equ. 4.93).

Also equivalent to the Verlet algorithm is the *velocity Verlet* prescription introduced by Swope [SWOPE 82]. The first line in Figure 4.6 looks like a simple Euler-Cauchy formula, but this is mere appearance. The quantity $v_{n+1}$ is *not* computed according to $v_{n+1} = v_n + b_n\Delta t$, as the EC method would require.

## 4.2.2   Predictor-Corrector Method

In the equation $d^2y/dt^2 = b(t)$ we again replace the function $b(t)$ by a NGB polynomial (see Sec. 4.1.5). Integrating twice, we obtain the general predictor formulae

$$\dot{y}^P_{n+1}\Delta t - \dot{y}_n\Delta t = (\Delta t)^2 \left[ b_n + \frac{1}{2}\nabla b_n + \frac{5}{12}\nabla^2 b_n + \frac{3}{8}\nabla^3 b_n + \right.$$
$$\left. +\frac{251}{720}\nabla^4 b_n + \frac{95}{288}\nabla^5 b_n + \dots \right] \qquad (4.99)$$

$$y^P_{n+1} - y_n - \dot{y}_n\Delta t = \frac{(\Delta t)^2}{2} \left[ b_n + \frac{1}{3}\nabla b_n + \frac{1}{4}\nabla^2 b_n + \frac{19}{90}\nabla^3 b_n + \right.$$
$$\left. +\frac{3}{16}\nabla^4 b_n + \frac{863}{5040}\nabla^5 b_n + \dots \right] \qquad (4.100)$$

A specific predictor of order $k$ is found by using terms up to order $\nabla^{k-2} b_n$. Thus the predictor of third order reads

$$\dot{y}^P_{n+1}\Delta t - \dot{y}_n\Delta t = (\Delta t)^2 \left[ \frac{3}{2}b_n - \frac{1}{2}b_{n-1} \right] + O[(\Delta t)^4] \qquad (4.101)$$

$$y^P_{n+1} - y_n - \dot{y}_n\Delta t = \frac{(\Delta t)^2}{2} \left[ \frac{4}{3}b_n - \frac{1}{3}b_{n-1} \right] + O[(\Delta t)^4] \qquad (4.102)$$

For a compact notation we define the vector

$$\mathbf{b}_k \equiv \{b_n, b_{n-1}, \dots b_{n-k+2}\}^T \qquad (4.103)$$

and the coefficient vectors $\mathbf{c}_k$ and $\mathbf{d}_k$. The predictor of order $k$ may then be written as

---

**Predictor of order k for second order DE:**

$$\dot{y}^P_{n+1}\Delta t - \dot{y}_n\Delta t = (\Delta t)^2 \mathbf{c}_k \cdot \mathbf{b}_k + O[(\Delta t)^{k+1}] \qquad (4.104)$$

$$y^P_{n+1} - y_n - \dot{y}_n\Delta t = \frac{(\Delta t)^2}{2}\mathbf{d}_k \cdot \mathbf{b}_k + O[(\Delta t)^{k+1}] \qquad (4.105)$$

---

The first few vectors $\mathbf{c}_k, \mathbf{d}_k$ are given by

$$\mathbf{c}_2 = 1 \qquad\qquad\qquad \mathbf{d}_2 = 1 \qquad\qquad (4.106)$$

$$\mathbf{c}_3 = \begin{pmatrix} 3/2 \\ -1/2 \end{pmatrix} \qquad \mathbf{d}_3 = \begin{pmatrix} 4/3 \\ -1/3 \end{pmatrix} \qquad (4.107)$$

$$\mathbf{c}_4 = \begin{pmatrix} 23/12 \\ -16/12 \\ 5/12 \end{pmatrix} \qquad \mathbf{d}_4 = \begin{pmatrix} 19/12 \\ -10/12 \\ 3/12 \end{pmatrix} \qquad (4.108)$$

$$\mathbf{c}_5 = \begin{pmatrix} 55/24 \\ -59/24 \\ 37/24 \\ -9/24 \end{pmatrix} \qquad \mathbf{d}_5 = \begin{pmatrix} 323/180 \\ -264/180 \\ 159/180 \\ -38/180 \end{pmatrix} \tag{4.109}$$

Having performed the predictor step, we may insert the preliminary result $y_{n+1}^P, \dot{y}_{n+1}^P$ in the physical law for $b[y, \dot{y}]$. This *evaluation step* yields

$$b_{n+1}^P \equiv b \left[ y_{n+1}^P, \dot{y}_{n+1}^P \right] \tag{4.110}$$

(If the acceleration $b$ is effected by a potential force that depends on $y$ but not on $\dot{y}$, the quantity $\dot{y}_{n+1}^P$ need not be computed at all.) By inserting $b_{n+1}^P$ now in a NGB formula centered on $t_{n+1}$ and again integrating twice we find the general corrector

$$\dot{y}_{n+1}\Delta t - \dot{y}_n \Delta t = (\Delta t)^2 \left[ b_{n+1}^P - \frac{1}{2} \nabla b_{n+1} - \frac{1}{12} \nabla^2 b_{n+1} - \frac{1}{24} \nabla^3 b_{n+1} - \right.$$

$$\left. - \frac{19}{720} \nabla^4 b_{n+1} - \frac{3}{160} \nabla^5 b_{n+1} - \cdots \right] \tag{4.111}$$

$$y_{n+1} - y_n - \dot{y}_n \Delta t = \frac{(\Delta t)^2}{2} \left[ b_{n+1}^P - \frac{2}{3} \nabla b_{n+1} - \frac{1}{12} \nabla^2 b_{n+1} - \frac{7}{180} \nabla^3 b_{n+1} - \right.$$

$$\left. - \frac{17}{720} \nabla^4 b_{n+1} - \frac{41}{2520} \nabla^5 b_{n+1} - \cdots \right] \tag{4.112}$$

Defining the vector

$$\mathbf{b}_k^P \equiv \{ b_{n+1}^P, b_n, \ldots b_{n-k+3} \}^T \tag{4.113}$$

and another set of coefficient vectors $\mathbf{e}_k, \mathbf{f}_k$, we may write the corrector of order $k$ as

---

**Corrector of order k for second-order DE:**

$$\dot{y}_{n+1}\Delta t - \dot{y}_n \Delta t = (\Delta t)^2 \mathbf{e}_k \cdot \mathbf{b}_k^P + O[(\Delta t)^{k+1}] \tag{4.114}$$

$$y_{n+1} - y_n - \dot{y}_n \Delta t = \frac{(\Delta t)^2}{2} \mathbf{f}_k \cdot \mathbf{b}_k^P + O[(\Delta t)^{k+1}] \tag{4.115}$$

---

The first few coefficient vectors are

$$\mathbf{e}_2 = 1 \qquad\qquad \mathbf{f}_2 = 1 \tag{4.116}$$

$$\mathbf{e}_3 = \begin{pmatrix} 1/2 \\ 1/2 \end{pmatrix} \qquad\qquad \mathbf{f}_3 = \begin{pmatrix} 1/3 \\ 2/3 \end{pmatrix} \tag{4.117}$$

$$\mathbf{e}_4 = \begin{pmatrix} 5/12 \\ 8/12 \\ -1/12 \end{pmatrix} \qquad \mathbf{f}_4 = \begin{pmatrix} 3/12 \\ 10/12 \\ -1/12 \end{pmatrix} \tag{4.118}$$

$$\mathbf{e}_5 = \begin{pmatrix} 9/24 \\ 19/24 \\ -5/24 \\ 1/24 \end{pmatrix} \qquad \mathbf{f}_5 = \begin{pmatrix} 38/180 \\ 171/180 \\ -36/180 \\ 7/180 \end{pmatrix} \tag{4.119}$$

The PC method should always be applied according to the scheme P(EC)E. Repeating the corrector step, as in $P(EC)^2E$, is uneconomical. Of course, omitting the corrector step altogether is not to be recommended either. The bare predictor scheme PE is tantamount to using one of the *explicit* algorithms whose bad stability rating we have discussed in Section 4.1.3.

### 4.2.3 Nordsieck Formulation of the PC Method

There are two ways of extrapolating a function – as, for instance, the solution $y(t)$ of our differential equation – into the time interval $[t_n, t_{n+1}]$. One is to thread a NGB polynomial through a number of preceding points $\{t_{n-k}, y_{n-k}\}$; the other is to write down a Taylor expansion about $t_n$. For the latter approach one needs, instead of the stored values $y_{n-k}$, a few *derivatives* $d^k y/dt^k$ at $t_n$. Such a Taylor predictor of order 3 would read

$$y_{n+1}^P = y_n + \dot{y}_n \Delta t + \ddot{y}_n \frac{(\Delta t)^2}{2!} + \dddot{y}_n \frac{(\Delta t)^3}{3!} + O[(\Delta t)^4] \tag{4.120}$$

$$\dot{y}_{n+1}^P \Delta t = \dot{y}_n \Delta t + \ddot{y}_n (\Delta t)^2 + \dddot{y}_n \frac{(\Delta t)^3}{2!} + O[(\Delta t)^4] \tag{4.121}$$

$$\ddot{y}_{n+1}^P \frac{(\Delta t)^2}{2!} = \ddot{y}_n \frac{(\Delta t)^2}{2!} + \dddot{y}_n \frac{(\Delta t)^3}{2!} + O[(\Delta t)^4] \tag{4.122}$$

$$\dddot{y}_{n+1}^P \frac{(\Delta t)^3}{3!} = \dddot{y}_n \frac{(\Delta t)^3}{3!} + O[(\Delta t)^4] \tag{4.123}$$

Defining the vector

$$\mathbf{z}_n \equiv \begin{pmatrix} y_n \\ \dot{y}_n \Delta t \\ \ddot{y}_n \frac{(\Delta t)^2}{2!} \\ \vdots \end{pmatrix} \tag{4.124}$$

and the (Pascal triangle) matrix

$$\mathbf{A} \equiv \begin{pmatrix} 1 & 1 & 1 & 1 & \cdots \\ 0 & 1 & 2 & 3 & \cdots \\ 0 & 0 & 1 & 3 & \cdots \\ & \ddots & & \ddots & 1 & \ddots \\ & & & & & \ddots \end{pmatrix} \tag{4.125}$$

we have

$$\boxed{z_{n+1}^P = \mathbf{A} \cdot z_n} \tag{4.126}$$

Now follows the evaluation step. Inserting the relevant components of $z_{n+1}^P$ in the given force law we obtain the acceleration

$$b_{n+1}^P \equiv b[y_{n+1}^P, \dot{y}_{n+1}^P], \tag{4.127}$$

which in general will deviate from the extrapolated acceleration as given by equ. 4.122. We define a correction term

$$\gamma \equiv [b_{n+1}^P - \ddot{y}_{n+1}^P]\frac{(\Delta t)^2}{2} \tag{4.128}$$

and write the *corrector* for $z_{n+1}$ as

$$\boxed{z_{n+1} = z_{n+1}^P + \gamma \mathbf{c}} \tag{4.129}$$

with an optimized coefficient vector $\mathbf{c}$ [GEAR 66]. For the first few orders of accuracy this vector is given as

$$\mathbf{c} = \begin{pmatrix} 1/6 \\ 5/6 \\ 1 \\ 1/3 \end{pmatrix}, \begin{pmatrix} 19/120 \\ 3/4 \\ 1 \\ 1/2 \\ 1/12 \end{pmatrix}, \begin{pmatrix} 3/20 \\ 251/360 \\ 1 \\ 11/18 \\ 1/6 \\ 1/60 \end{pmatrix}, \dots \tag{4.130}$$

These coefficients were optimized by Gear under the assumption that $b$ depends on the position coordinate $y$ only, being independent of $\dot{y}$. The simple but important case of point masses interacting via potential forces is covered by this apparatus. Whenever $b = b(y, \dot{y})$, as in rigid body rotation or for velocity dependent forces, Gear recommends to replace $19/120$ by $19/90$ and $3/20$ by $3/16$ (see Appendix C of [GEAR 66]).

Finally, the evaluation step is repeated to yield an improved value of the acceleration, $b_{n+1}$. As before, the procedure may be described in short notation as P(EC)E.

The Nordsieck PC method offers the advantage of being self-starting – provided that one adds to the initial conditions $y_0, \dot{y}_0$ and the corresponding acceleration $\ddot{y}_0$ some ad hoc assumptions about the values of $\dddot{y}_0, \ddddot{y}_0$ etc. (for instance, $\dots = 0$). As in all self-starting (single step) algorithms it is possible to modify the time step whenever necessary.

Stability analysis is somewhat tedious for this formulation of the PC method, but there are no real surprises. Once again the quasi-implicit nature of the corrector provides a welcome extension of the stability region as compared to the bare predictor formula.

**Runge-Kutta scheme of 4th order for second order DE:**

$$b_1 = b[y_n]$$

$$b_2 = b\left[y_n + \dot{y}_n\frac{\Delta t}{2}\right]$$

$$b_3 = b\left[y_n + \dot{y}_n\frac{\Delta t}{2} + b_1\frac{(\Delta t)^2}{4}\right]$$

$$b_4 = b\left[y_n + \dot{y}_n\Delta t + b_2\frac{(\Delta t)^2}{2}\right]$$

$$\dot{y}_{n+1} = \dot{y}_n + \frac{\Delta t}{6}[b_1 + 2b_2 + 2b_3 + b_4] + O[(\Delta t)^5] \tag{4.131}$$

$$y_{n+1} = y_n + \dot{y}_n\Delta t + \frac{(\Delta t)^2}{6}[b_1 + b_2 + b_3] + O[(\Delta t)^5] \tag{4.132}$$

Figure 4.7: Runge-Kutta algorithm of 4th order for a second-order DE with $b = b(y)$. (The coefficient $(\Delta t)^2/4$ in the expression for $b_3$ is correct; it has a different origin than the respective coefficient in Fig. 4.8 below.)

### 4.2.4  Runge-Kutta Method

The basic idea of the RK method was sketched in 4.1.6. Without giving a detailed derivation, we here list a widely used RK algorithm of fourth order for the equation $d^2y/dt^2 = b[y(t)]$ (see Figure 4.7). If the acceleration $b$ depends not only on $y$ but also on $\dot{y}$, then the procedure given in Figure 4.8 should be used [ABRAMOWITZ 65]. With regard to the economy of the RK method the considerations of Sec. 4.1.6 hold: the repeated evaluation of the acceleration $b(y)$ in the course of a single time step may be critical if that evaluation consumes much computer time; this more or less rules out the method for application in $N$-body simulations. In all other applications the RK method is usually the first choice. It is a self-starting algorithm, very accurate, and the assessment of the local truncation error using divided time steps is always possible.

### 4.2.5  Symplectic Algorithms

There is more to life than accuracy and stability. In recent years a class of integration schemes called "Hamiltonian" or "symplectic" algorithms have been discussed a lot. These are integration procedures that are particularly well suited for the treatment of mechanical equations of motion.

"Symplectic" means "interlaced" or "intertwined". The term, which is due to H. Weyl (cited in [GOLDSTEIN 80]), refers to a particular formulation of the classical Hamiltonian equations of motion. The motivation for the development

---

**Runge-Kutta scheme of 4th order for velocity dependent forces:**

$$b_1 = b[y_n, \dot{y}_n]$$

$$b_2 = b\left[y_n + \dot{y}_n\frac{\Delta t}{2} + b_1\frac{(\Delta t)^2}{8}, \dot{y}_n + b_1\frac{\Delta t}{2}\right]$$

$$b_3 = b\left[y_n + \dot{y}_n\frac{\Delta t}{2} + b_1\frac{(\Delta t)^2}{8}, \dot{y}_n + b_2\frac{\Delta t}{2}\right]$$

$$b_4 = b\left[y_n + \dot{y}_n\Delta t + b_3\frac{(\Delta t)^2}{2}, \dot{y}_n + b_3\Delta t\right]$$

$$\dot{y}_{n+1} = \dot{y}_n + \frac{\Delta t}{6}[b_1 + 2b_2 + 2b_3 + b_4] + O[(\Delta t)^5] \qquad (4.133)$$

$$y_{n+1} = y_n + \dot{y}_n\Delta t + \frac{(\Delta t)^2}{6}[b_1 + b_2 + b_3] + O[(\Delta t)^5] \qquad (4.134)$$

---

Figure 4.8: Runge-Kutta of 4th order for second-order DE with $b = b(y, \dot{y})$

of symplectic algorithms was the hope to "catch" the inherent characteristics of mechanical systems more faithfully than by indiscriminately applying one of the available integration schemes.

Consider a classical system with $M$ degrees of freedom. The complete set of (generalized) coordinates is denoted by $\mathbf{q}$, the conjugate momenta are called $\mathbf{p}$. Hamilton's equations read

$$\frac{d\mathbf{q}}{dt} = \nabla_p H(\mathbf{q}, \mathbf{p}) \qquad \frac{d\mathbf{p}}{dt} = -\nabla_q H(\mathbf{q}, \mathbf{p}) \qquad (4.135)$$

where $H(\mathbf{q}, \mathbf{p})$ is the (time-independent) Hamiltonian. By linking together the two $M$-vectors $\mathbf{q}$ and $\mathbf{p}$ we obtain a phase space vector $\mathbf{z}$ whose temporal evolution is described by the concise equation of motion

$$\frac{d\mathbf{z}}{dt} = \mathbf{J} \cdot \nabla_z H(\mathbf{z}) \qquad (4.136)$$

with the "symplectic matrix"

$$\mathbf{J} \equiv \begin{pmatrix} \mathbf{0} & \mathbf{I} \\ -\mathbf{I} & \mathbf{0} \end{pmatrix} \qquad (4.137)$$

A glance at this matrix makes the significance of the term "intertwined" apparent.

Let us now assume that we are to solve the dynamic equations with given initial conditions. If there is an exact solution, yielding $\mathbf{z}(t)$ from the initial vector $\mathbf{z}(t_0)$, the mapping

$$\mathbf{z}(t_0) \implies \mathbf{z}(t) \qquad (4.138)$$

represents a *canonical transformation* in phase space. It is well known that such a transformation conserves the energy (= numerical value of the Hamiltonian), and this property is often used to assess the quality of numerical approximations to the exact solution. However, there is another conserved quantity which has for a long time been disregarded as a measure of quality of numerical integrators. Namely, canonical transformations leave the *symplectic form*

$$s(\mathbf{z}_1, \mathbf{z}_2) \equiv \mathbf{z}_1^T \cdot \mathbf{J} \cdot \mathbf{z}_2 \tag{4.139}$$

unchanged. This tells us something about the "natural" evolution of volume elements (or rather, "bundles" of trajectories) in phase space. Indeed, Liouville's theorem, that (deterministic) cornerstone of statistical mechanics, follows from the conservation of the standard symplectic form.

<u>EXAMPLE:</u> Let us unclamp that harmonic oscillator once more. Writing, in honor of R. Hamilton, $q$ for the position and $p$ for the (conjugate) momentum, we have

$$H(\mathbf{z}) \equiv H(q, p) = \frac{k}{2}q^2 + \frac{p^2}{2m} \tag{4.140}$$

The canonical transformation producing the solution at time $t$ from the initial conditions $q(0), p(0)$ may be written

$$\mathbf{z}(t) = \begin{pmatrix} q \\ p \end{pmatrix} = \begin{pmatrix} \cos\omega t & \frac{1}{m\omega}\sin\omega t \\ -m\omega\sin\omega t & \cos\omega t \end{pmatrix} \cdot \begin{pmatrix} q(0) \\ p(0) \end{pmatrix} \equiv \mathbf{A} \cdot \mathbf{z}(0) \tag{4.141}$$

(with $\omega^2 = k/m$.) The energy is, of course, conserved:

$$\frac{k}{2}q^2 + \frac{p^2}{2m} = \frac{k}{2}q^2(0) + \frac{p^2(0)}{2m} \tag{4.142}$$

What about symplectic structure? Writing $\{q_1(0), p_1(0)\}$ and $\{q_2(0), p_2(0)\}$ for two different initial conditions we find

$$s(q_1(0) \ldots p_2(0)) \equiv (q_1(0), p_1(0)) \cdot \begin{pmatrix} 0 & 1 \\ -1 & 0 \end{pmatrix} \cdot \begin{pmatrix} q_2(0) \\ p_2(0) \end{pmatrix} \tag{4.143}$$

$$= q_1(0)p_2(0) - p_1(0)q_2(0) \tag{4.144}$$

There is a simple geometric interpretation for $s$. Regarding $\mathbf{z} \equiv \{q, p\}$ as a vector in two-dimensional phase space we see that $s$ is just the area of a parallelogram defined by the two initial state vectors $\mathbf{z}_{1,2}$. Let us check whether $s$ is constant under the transformation 4.141:

$$s(\mathbf{z}_1(t), \mathbf{z}_2(t)) = \mathbf{z}_1^T(t) \cdot \mathbf{J} \cdot \mathbf{z}_2(t) \tag{4.145}$$

$$= \mathbf{z}_1^T(0) \cdot \mathbf{A}^T \cdot \mathbf{J} \cdot \mathbf{A} \cdot \mathbf{z}_2(0) \tag{4.146}$$

$$= \mathbf{z}_1^T(0) \cdot \mathbf{J} \cdot \mathbf{z}_2(0) \tag{4.147}$$

In other words, the matrices $\mathbf{A}$ and $\mathbf{J}$ fulfill the requirement $\mathbf{A}^T \cdot \mathbf{J} \cdot \mathbf{A} = \mathbf{J}$.

So much for the exact solution. Now for the simplest numerical integrator, the Euler-Cauchy scheme. It may be written as

$$\mathbf{z}(t) = \begin{pmatrix} q \\ p \end{pmatrix} = \begin{pmatrix} 1 & \frac{\Delta t}{m} \\ -m\omega^2 \Delta t & 1 \end{pmatrix} \cdot \begin{pmatrix} q(0) \\ p(0) \end{pmatrix} \equiv \mathbf{E} \cdot \mathbf{z}(0) \qquad (4.148)$$

It is easy to prove that this procedure enhances both the energy and the symplectic form by a factor $1+(\omega\Delta t)^2$ at each time step. In this simple case there is an easy remedy: dividing the Euler-Cauchy matrix by $\sqrt{1+(\omega\Delta t)^2}$ we obtain an integrator that conserves both the energy and the symplectic structure exactly. Of course, this is just a particularly harmonious feature of our domestic oscillator.

There are several ways of constructing symplectic algorithms. After pioneering attempts by various groups the dust has settled a bit, and the very readable survey paper by Yoshida provides a good overview, with all important citations [YOSHIDA 93].

A symplectic integrator of fourth order that has been developed independently by Neri and by Candy and Rozmus is described in Fig. 4.9 [NERI 88], [CANDY 91].

Note that the Candy algorithm is *explicit* and resembles a Runge-Kutta procedure; in contrast to a fourth-order RK algorithm is requires only three force evaluations per time step. A third-order scheme (comparable in accuracy to Størmer-Verlet) was found by R. D. Ruth; it has the same structure as Candy's algorithm, with the coefficients [RUTH 83]

$$\begin{aligned} (a_1, a_2, a_3) &= (2/3, -2/3, 1) & (4.151) \\ (b_1, b_2, b_3) &= (7/24, 3/4, -1/24) & (4.152) \end{aligned}$$

For Hamiltonians that are not separable with respect to $\mathbf{q}$ and $\mathbf{p}$ symplectic algorithms may be devised as well. However, they must be *implicit* schemes [YOSHIDA 93].

Of course, the various time-proven algorithms discussed in the preceding sections have all been examined for their symplecticity properties. Only one among them conserves symplectic structure: the Størmer-Verlet formula. The venerable Runge-Kutta scheme fails, and so do the PC methods.

Is it not an unprofitable enterprise to construct an integrator that conserves so seemingly abstract a quantity as $s(\mathbf{z}_1, \mathbf{z}_2)$? Not quite. It is a well-established fact that for non-integrable Hamiltonians (and as one might guess, practically all interesting systems are non-integrable) there can be no algorithm that conserves *both* energy and symplectic structure. But Yoshida has shown that symplectic integrators do conserve a Hamiltonian function that is different from, *but close to*, the given Hamiltonian [YOSHIDA 93]. As a consequence, symplectic algorithms will display no secular (i.e. long-time) growth of error with regard to energy. This is in marked contrast to the behavior of, say, the usual Runge-Kutta integrators,

**Symplectic algorithm of fourth order:** Let the Hamiltonian be separable in terms of coordinates and momenta: $H(\mathbf{q}, \mathbf{p}) = U(\mathbf{q}) + T(\mathbf{p})$. For the derivatives of $H$ we use the notation

$$\mathbf{F}(\mathbf{q}) \equiv -\nabla_q U(\mathbf{q}), \quad \mathbf{P}(\mathbf{q}) \equiv \nabla_p T(\mathbf{p}) \qquad (4.149)$$

The state at time $t$ is given by $\{\mathbf{q}_0, \mathbf{p}_0\}$.

- For $i = 1$ to 4 do

$$\mathbf{p}_i = \mathbf{p}_{i-1} + b_i \mathbf{F}(\mathbf{q}_{i-1})\Delta t, \quad \mathbf{q}_i = \mathbf{q}_{i-1} + a_i \mathbf{P}(\mathbf{p}_i)\Delta t \qquad (4.150)$$

  where

$$
\begin{aligned}
a_1 = a_4 &= (2 + 2^{1/3} + 2^{-1/3})/6 \\
a_2 = a_3 &= (1 - 2^{1/3} - 2^{-1/3})/6 \\
b_1 &= 0 \\
b_2 = b_4 &= 1/(2 - 2^{1/3}) \\
b_3 &= 1/(1 - 2^{2/3})
\end{aligned}
$$

- The state at time $t_{n+1}$ is $\{\mathbf{q}_4, \mathbf{p}_4\}$.

Figure 4.9: Symplectic algorithm by Neri and Candy

which show good local (short-time) accuracy but when applied to Hamiltonian systems will lead to a regularly increasing deviation in energy.

To be specific, the simple first-order symplectic algorithm

$$\mathbf{p}_{n+1} = \mathbf{p}_n + \mathbf{F}(\mathbf{q}_n)\Delta t, \quad \mathbf{q}_{n+1} = \mathbf{q}_n + \mathbf{P}(\mathbf{p}_{n+1})\Delta t \qquad (4.153)$$

exactly conserves a Hamiltonian $\tilde{H}$ that is associated to the given Hamiltonian $H$ by

$$\tilde{H} \equiv H + H_1\Delta t + H_2(\Delta t)^2 + H_3(\Delta t)^3 + \dots \qquad (4.154)$$

where

$$H_1 = \frac{1}{2}H_p H_q, \quad H_2 = \frac{1}{12}(H_{pp}H_q^2 + H_{qq}H_p^2), \quad H_3 = \frac{1}{12}H_{pp}H_{qq}H_pH_q \ \dots \ (4.155)$$

($H_q$ being shorthand for $\nabla_q H$ etc.) In particular, for the harmonic oscillator the perturbed Hamiltonian

$$\tilde{H} = H_{ho} + \frac{\omega^2\Delta t}{2}pq \qquad (4.156)$$

is conserved exactly.

Incidentally, the one-step algorithm 4.153 is also known as the *Euler-Cromer method*. When applied to oscillator-like equations of motion it is a definite improvement over the (unstable) Euler-Cauchy method of equ. 4.7.

EXERCISE: Apply the (non-symplectic) RK method and the (symplectic) Størmer-Verlet algorithm (or the Candy procedure) to the one-body Kepler problem with elliptic orbit. Perform long runs to assess the long-time performance of the integrators. (For RK the orbit should eventually spiral down towards the central mass, while the symplectic procedures should only give rise to a gradual precession of the perihelion.)

## 4.2.6    Numerov's Method

This technique is usually discussed in the context of *boundary value problems* (BVP), although it is really an algorithm designed for use with a specific *initial value problem* (IVP). The reason is that in the framework of the so-called *shooting method* the solution to a certain kind of BVP is found by taking a detour over a related IVP (see Sec. 4.3.1). An important class of BVP has the general form

$$\frac{d^2y}{dx^2} = -g(x)y + s(x) \qquad (4.157)$$

with given boundary values $y(x_1)$ and $y(x_2)$. A familiar example is the one-dimensional Poisson equation for the potential $\phi(x)$ in the presence of a charge density $\rho(x)$,

$$\frac{d^2\phi}{dx^2} = -\rho(x) \qquad (4.158)$$

with the values of $\phi$ being given at $x_1$ and $x_2$. In terms of equ. 4.157, $g(x) = 0$ and $s(x) = -\rho(x)$.

The *shooting method* then consists in temporarily omitting the information $y(x_2)$, replacing it by a suitably estimated derivative $y'$ at $x_1$ and solving the initial value problem defined by $\{y(x_1), y'(x_1)\}$ – for example, by the Numerov method. By comparing the end value of $y(x_2)$ thus computed to the given boundary value at $x_2$ one may systematically improve $y'(x_1)$, approaching the correct solution in an iterative manner.

To implement Numerov's method one divides the interval $[x_1, x_2]$ into subintervals of length $\Delta x$ and at each intermediate point $x_n$ expands $y(x)$ into a power series. Adding the Taylor formulae for $y_{n+1}$ and $y_{n-1}$ one finds

$$y_{n+1} = 2y_n - y_{n-1} + y_n''(\Delta x)^2 + y_n^{(4)}\frac{(\Delta x)^4}{12} + O[(\Delta x)^6] \qquad (4.159)$$

(Note that up to the third term on the r.h.s. this is just Verlet's formula 4.85.) Insertion of the specific form 4.157 of $y_n''$ yields

$$y_{n+1} = 2y_n - y_{n-1} + (\Delta x)^2[-g_n y_n + s_n] + \frac{(\Delta x)^4}{12}y_n^{(4)} + O[(\Delta x)^6] \qquad (4.160)$$

For the fourth derivative $y^{(4)}$ one writes, to the same order of accuracy,

$$\begin{aligned}
y_n^{(4)} &= \left.\frac{d^2 y''}{dx^2}\right|_n = \left.\frac{d^2(-gy+s)}{dx^2}\right|_n \approx \frac{1}{(\Delta x)^2}\delta_n^2(-gy+s) = \\
&= \frac{1}{(\Delta x)^2}[-g_{n+1}y_{n+1} + 2g_n y_n - g_{n-1}y_{n-1} + \\
&\qquad\qquad\qquad\qquad\qquad +s_{n+1} - 2s_n + s_{n-1}] \qquad (4.161)
\end{aligned}$$

Inserting this in 4.160 one arrives at Numerov's formula

$$\begin{aligned}
y_{n+1}[1 + \frac{(\Delta x)^2}{12}g_{n+1}] &= 2y_n[1 - \frac{5}{12}(\Delta x)^2 g_n] - y_{n-1}[1 + \frac{(\Delta x)^2}{12}g_{n-1}] + \\
&\quad + \frac{(\Delta x)^2}{12}[s_{n+1} + 10s_n + s_{n-1}] + O[(\Delta x)^6] \qquad (4.162)
\end{aligned}$$

To start this two-step algorithm at the point $x_1$ one needs an estimated value of $y(x_1 - \Delta x)$. Alternatively, one may estimate $y'(x_1)$ and treat the first subinterval by some self-starting single step algorithm such as Runge-Kutta.

EXERCISE: Write a code that permits to solve a given second-order equation of motion by various algorithms. Apply the program to problems of point mechanics and explore the stabilities and accuracies of the diverse techniques.

## 4.3    Boundary Value Problems

The general form of a BVP with one independent variable is

$$\frac{dy_i}{dx} = f_i(x, y_1, \ldots y_N) ; \quad i = 1, \ldots N \tag{4.163}$$

where the $N$ required boundary values are now given at more than one point $x$. Typically there are

$n_1$    boundary values $a_j$ $(j = 1, \ldots n_1)$ at $x = x_1$, and
$n_2$    $\equiv N - n_1$ boundary values $b_k$ $(k = 1, \ldots n_2)$ at $x = x_2$.

Of course, the quantities $y_i, a_j$ and $b_k$ may simply be higher derivatives of a single solution function $y(x)$. In physics we often encounter BVPs of the type

$$\frac{d^2 y}{dx^2} = -g(x)y + s(x) \tag{4.164}$$

which may be transformed, via the substitutions $y_1 \equiv y$, $y_2 \equiv -g(x)y_1 + s(x)$, into

$$\frac{dy_1}{dx} = y_2 \tag{4.165}$$

$$\frac{dy_2}{dx} = -g(x)y_1 + s(x) \tag{4.166}$$

Important examples of this kind of boundary value problems are Poisson's and Laplace's equations and the time independent Schroedinger equation.

The one-dimensional Poisson equation reads $d^2\phi/dx^2 = -\rho(x)$, or

$$\frac{d\phi}{dx} = -e \tag{4.167}$$

$$\frac{de}{dx} = \rho(x) \tag{4.168}$$

where $\rho(x)$ is a charge density. Laplace's equation is identical to Poisson's, but with $\rho(x) = 0$, i.e. in charge-free space. Another physical problem described by the same equation is the temperature distribution along a thin rod: $d^2T/dx^2 = 0$.

The Schroedinger equation for a particle of mass $m$ in a potential $U(x)$ reads

$$\frac{d^2\psi}{dx^2} = -g(x)\psi, \quad \text{with } g(x) = \frac{2m}{\hbar^2}[E - U(x)] \tag{4.169}$$

Also, the case of a particle on a centrosymmetric potential $U(r)$ may be treated by the same formalism. Factorizing the wave function as in

$$\psi(\mathbf{r}) \equiv \frac{1}{r} R(r) Y_{lm}(\theta, \phi) \tag{4.170}$$

we have for the radial function $R(r)$

$$\frac{d^2 R}{dr^2} = -g(r)R, \qquad (4.171)$$

$$\text{with} \quad g(r) = \frac{2m}{\hbar^2}\left[E - U(r) - \frac{l(l+1)\hbar^2}{2mr^2}\right] \qquad (4.172)$$

Two methods are available for finding a solution to any boundary value problem, not necessarily of the form 4.164. They are known as the *shooting* and the *relaxation* technique, respectively.

## 4.3.1   Shooting Method

The basic strategy here is to transform the given *boundary* value problem into an *initial* value problem with estimated parameters that are then iteratively adjusted so as to reproduce the given boundary values. The detailed procedure is as follows:

**First trial shot:** Augment the $n_1$ boundary values given at $x = x_1$ by $n_2 \equiv N - n_1$ *estimated* parameters

$$\mathbf{a}^{(1)} \equiv \{a_k^{(1)}; \ k = 1, \ldots n_2\}^T \qquad (4.173)$$

such that a completely determined initial value problem is obtained. Now integrate this IVP by some suitable technique up to the second boundary point $x = x_2$. (For equations of the frequently occuring form $y'' = -g(x)y + s(x)$ Numerov's method is recommended.) The newly calculated functional values at $x = x_2$,

$$\mathbf{b}^{(1)} \equiv \{b_k^{(1)}; \ k = 1, \ldots n_2\}^T \qquad (4.174)$$

will in general deviate from the given boundary values $\mathbf{b} \equiv \{b_k; \ldots\}^T$. The difference vector

$$\mathbf{e}^{(1)} \equiv \mathbf{b}^{(1)} - \mathbf{b} \qquad (4.175)$$

is stored for further use.

**Second trial shot:** Change the estimated initial values $a_k$ by some small amount:

$$\mathbf{a}^{(2)} \equiv \mathbf{a}^{(1)} + \delta\mathbf{a} \qquad (4.176)$$

and again perform the integration up to $x = x_2$. The boundary values $b_k^{(2)}$ thus obtained are again different from the required values $b_k$:

$$\mathbf{e}^{(2)} \equiv \mathbf{b}^{(2)} - \mathbf{b} \qquad (4.177)$$

**Quasi-linearization:** Assuming that the deviations $\mathbf{e}^{(1)}$ and $\mathbf{e}^{(2)}$ depend *linearly* on the estimated initial values $\mathbf{a}^{(1)}$ and $\mathbf{a}^{(2)}$, we may compute that vector $\mathbf{a}^{(3)}$ which would make the deviations disappear (Newton-Raphson technique):

$$\mathbf{a}^{(3)} = \mathbf{a}^{(1)} - \mathbf{A}^{-1} \cdot \mathbf{e}^{(1)}, \quad \text{with} \ A_{ij} \equiv \frac{b_i^{(2)} - b_i^{(1)}}{a_j^{(2)} - a_j^{(1)}} \qquad (4.178)$$

As a rule the vectors **e** are in fact not exactly linear in **a**. Therefore one has
to iterate the procedure, putting $\mathbf{a}^{(1)} = \mathbf{a}^{(2)}$ and $\mathbf{a}^{(2)} = \mathbf{a}^{(3)}$ etc., until some
desired accuracy has been achieved.

EXAMPLE: Let the boundary value problem be defined by the DE

$$\frac{d^2y}{dx^2} = -\frac{1}{(1+y)^2}$$  (4.179)

with given values $y(0) = y(1) = 0$.
*First trial shot:* To obtain a completely determined IVP, we choose $a^{(1)} \equiv y'(0) = 1.0$.
Application of a 4th order Runge-Kutta integrator with 10 sub-intervals $\Delta x = 0.1$ yields
$b^{(1)} \equiv y(1) = 0.674$. Since the required boundary value at $x = 1$ is $y(1) = 0$ the deviation
is $e^{(1)} = 0.674$.
*Second trial shot:* Now we put $a^{(2)} = 1.1$ and integrate once more, finding $b^{(2)} = 0.787$,
i.e. $e^{(2)} = 0.787$.
*Quasi-linearization:* From

$$a^{(3)} = a^{(1)} - \frac{a^{(2)} - a^{(1)}}{b^{(2)} - b^{(1)}} e^{(1)}$$  (4.180)

we find $a^{(3)} = 0.405 \,(\equiv y'(0))$.
*Iteration:* The next few iterations yield the following values for $a \,(\equiv y'(0))$ and $b \,(\equiv y(1))$:

| $n$ | $a^{(n)}$ | $b^{(n)}$ |
|---|---|---|
| 3 | 0.405 | − 0.041 |
| 4 | 0.440 | 0.003 |
| 5 | 0.437 | 0.000 |

It is sometimes inconvenient to integrate the (artificial) initial value problem
over the entire interval $[x_1, x_2]$. Physical conditions (forces, densities, etc.) may
vary in different sub-regions of that interval, so that different step sizes, or even
algorithms, are appropriate. In such cases one defines internal border points $x_b$
between such subintervals and joins the piecewise solution functions together by
requiring smooth continuation at $x_b$. An example for this variant of the shooting
method is given in [KOONIN 85].

## 4.3.2    Relaxation Method

By discretizing the independent variable $x$ we may always transform a given DE
into a set of algebraic equations. For example, in the equation

$$\frac{d^2y}{dx^2} = b(x, y)$$  (4.181)

the second derivative may be replaced by the DDST approximation

$$\frac{d^2y}{dx^2} \approx \frac{1}{(\Delta x)^2}[y_{i+1} - 2y_i + y_{i-1}] \tag{4.182}$$

which leads to the set of equations

$$y_{i+1} - 2y_i + y_{i-1} - b_i(\Delta x)^2 = 0, \quad i = 2, \ldots M-1 \tag{4.183}$$

The values of $y_1$ and $y_M$ will be given: we are dealing with a boundary value problem.

Assume now that we are in possession of a set of values $y_i$, compactly written as a vector $\mathbf{y}^{(1)}$, that solve the equations 4.183 approximately but not exactly. The error components

$$e_i = y_{i+1} - 2y_i + y_{i-1} - b_i(\Delta x)^2, \quad i = 2, \ldots M-1 \tag{4.184}$$

together with $e_1 = e_M = 0$ then define an error vector $\mathbf{e}^{(1)}$ which we want to make disappear by varying the components of $\mathbf{y}^{(1)}$. To find out what alterations in $\mathbf{y}^{(1)}$ will do the trick we expand the error components $e_i$ linearly in terms of the relevant $y_j$:

$$e_i(y_{i-1} + \Delta y_{i-1}, y_i + \Delta y_i, y_{i+1} + \Delta y_{i+1}) \approx$$
$$\approx e_i + \frac{\partial e_i}{\partial y_{i-1}}\Delta y_{i-1} + \frac{\partial e_i}{\partial y_i}\Delta y_i + \frac{\partial e_i}{\partial y_{i+1}}\Delta y_{i+1}$$
$$\equiv e_i + \alpha_i\Delta y_{i-1} + \beta_i\Delta y_i + \gamma_i\Delta y_{i+1} \quad (i = 1, \ldots M) \tag{4.185}$$

This modified error vector is called $\mathbf{e}^{(2)}$. The requirement $\mathbf{e}^{(2)} = 0$ may be written as

$$\mathbf{A} \cdot \Delta\mathbf{y} = -\mathbf{e}^{(1)} \tag{4.186}$$

with

$$\mathbf{A} = \begin{pmatrix} \beta_1 & \gamma_1 & 0 & \cdots \\ \alpha_2 & \beta_2 & \gamma_2 & 0 \\ & \ddots & \ddots & \ddots \\ & & \alpha_M & \beta_M \end{pmatrix} \tag{4.187}$$

(If $y(x_1)$ and $y(x_M)$ are given, then $\gamma_1 = \alpha_M = 0$ and $\beta_1 = \beta_M = 1$.) Thus our system of equations is tridiagonal and may readily be solved by the recursion technique of Section 2.1.4.

EXAMPLE: We take the same example as for the shooting method,

$$\frac{d^2y}{dx^2} = -\frac{1}{(1+y)^2} \tag{4.188}$$

with $y(0) = y(1) = 0$. The Stirling approximation to the second derivative yields

$$e_i = y_{i+1} - 2y_i + y_{i-1} + \frac{(\Delta x)^2}{(1+y_i)^2} \tag{4.189}$$

and thus

$$\alpha_i \equiv \frac{\partial e_i}{\partial y_{i-1}} = 1; \qquad \gamma_i \equiv \frac{\partial e_i}{\partial y_{i+1}} = 1;$$

$$\beta_i \equiv \frac{\partial e_i}{\partial y_i} = -2\left[1 + \frac{(\Delta x)^2}{(1+y_i)^3}\right] \qquad (4.190)$$

for $i = 2, \ldots M - 1$. Furthermore, we have $\alpha_1 = \gamma_1 = 0$, $\beta_1 = 1$ and $\alpha_M = \gamma_M = 0$, $\beta_M = 1$. Therefore we may write

$$\begin{pmatrix} 1 & 0 & 0 & \cdots \\ 1 & \beta_2 & 1 & 0 \\ 0 & \ddots & \ddots & \ddots \\ & & 0 & 1 \end{pmatrix} \cdot \begin{pmatrix} \Delta y_1 \\ \cdot \\ \cdot \\ \cdot \\ \Delta y_M \end{pmatrix} = -\begin{pmatrix} e_1 \\ \cdot \\ \cdot \\ \cdot \\ e_M \end{pmatrix} \qquad (4.191)$$

To start the downwards recursion we put $g_{M-1} = -\alpha_M/\beta_M = 0$ and $h_{M-1} = -e_M/\beta_M = 0$. The recursion

$$g_{i-1} = \frac{-\alpha_i}{\beta_i + \gamma_i g_i} = \frac{-1}{\beta_i + g_i}; \quad h_{i-1} = \frac{-e_i - h_i}{\beta_i + g_i} \qquad (4.192)$$

brings us down to $g_1, h_1$. Putting

$$\Delta y_1 = \frac{-e_1 - \gamma_1 h_1}{\beta_1 + \gamma_1 g_1} = e_1 (= 0) \qquad (4.193)$$

we take the upwards recursion

$$\Delta y_{i+1} = g_i \Delta y_i + h_i; \quad i = 1, \ldots M - 1 \qquad (4.194)$$

to find the corrections $\Delta y_i$. Improved values of $y_i$ are formed according to $y_i \longrightarrow y_i + \Delta y_i$ and inserted in 4.189. After a few iterations these corrections are negligible.

# Chapter 5

# Partial Differential Equations

*Waves: a hyperbolic-advective process*

Entering now the vast field of partial differential equations, we immediately announce that our discussion shall be restricted to those types of equations that are of major importance in physics. These are the *quasilinear PDEs of second order*, which may be written in the general form

$$a_{11}\frac{\partial^2 u}{\partial x^2} + 2a_{12}\frac{\partial^2 u}{\partial x \partial y} + a_{22}\frac{\partial^2 u}{\partial y^2} + f(x, y, u, \frac{\partial u}{\partial x}, \frac{\partial u}{\partial y}) = 0 \qquad (5.1)$$

("Quasilinear" means that the second derivatives of $u$ appear in linear order only).

The official typology of partial differential equations distinguishes three types of such equations, viz. *hyperbolic*, *parabolic*, and *elliptic*:

*hyperbolic:*   $a_{11}a_{22} - a_{12}^2 < 0$     (or in particular $a_{12} = 0$, $a_{11}a_{22} < 0$)
*parabolic:*   $a_{11}a_{22} - a_{12}^2 = 0$     (or $a_{12} = 0$, $a_{11}a_{22} = 0$)
*elliptic:*   $a_{11}a_{22} - a_{12}^2 > 0$     (or $a_{12} = 0$, $a_{11}a_{22} > 0$)

Table 5.1 lists a few important examples for these kinds of PDEs.

| hyperbolic | $c^2\dfrac{\partial^2 u}{\partial x^2} - \dfrac{\partial^2 u}{\partial t^2} = f(x,t)$ | Wave equation |
|---|---|---|
| | $c^2\dfrac{\partial^2 u}{\partial x^2} - \dfrac{\partial^2 u}{\partial t^2} - a\,\dfrac{\partial u}{\partial t} = f(x,t)$ | Wave with damping |
| parabolic | $D\dfrac{\partial^2 u}{\partial x^2} - \dfrac{\partial u}{\partial t} = f(x,t)$ | Diffusion equation |
| | $\dfrac{\hbar^2}{2m}\dfrac{\partial^2 u}{\partial x^2} + i\hbar\,\dfrac{\partial u}{\partial t} - U(x)\,u = 0$ | Schroedinger equation |
| elliptic | $\dfrac{\partial^2 u}{\partial x^2} + \dfrac{\partial^2 u}{\partial y^2} = -\rho(x,y)$ | Potential equation |
| | $\dfrac{\partial^2 u}{\partial x^2} + \dfrac{\partial^2 u}{\partial y^2} - \dfrac{2m}{\hbar^2}\,U(x)\,u = 0$ <br> (or $\ldots = \epsilon\,u$) | Schroedinger equation, <br> stationary case |

Table 5.1: Some PDEs (partial differential equations) in physics

In the context of physical theory *hyperbolic* and *parabolic* equations as a rule describe *initial value problems*, which is to say that one of the independent variables is the time $t$, and that for $t = 0$ the values of $u$ and $\partial u/\partial t$ are known throughout the spatial region under scrutiny. The reason for this state of affairs is that such equations arise naturally from the description of transport phenomena, i.e. time-dependent problems. In contrast, *elliptic* PDEs as a rule occur in the description of stationary states $u(x,y)$, the variables $x$ and $y$ (and possibly a third independent variable, $z$) being spatial coordinates. The values of the stationary function $u(x,y)$ must then be given along a boundary curve $C(x,y) = 0$ (or surface, $S(x,y,z) = 0$): we are dealing with a *boundary value problem*. With the usual "controlled sloppyness" of physicists in matters mathematical we write:

$$\left.\begin{array}{l} \text{hyperbolic} \\ \text{parabolic} \end{array}\right\} \quad \Longleftrightarrow \quad \text{initial value problems}$$

$$\text{elliptic} \qquad\qquad \Longleftrightarrow \quad \text{boundary value problems}$$

Furthermore, we will restrict the discussion of initial value problems (IVP) to certain "pure" types which do not exhaust the vast multitude of hyperbolic and parabolic PDEs. The equations that are relevant to the description of physical transport processes are usually derived under the additional assumption that the quantity to be transported (mass, energy, momentum, charge, etc.) is conserved as a whole. The resulting *law of continuity* leads to (hyperbolic or parabolic) equations which are called *conservative*.

Let the spatial distribution of some measurable quantity be described by a "density" $u(\mathbf{r}, t)$. Just for simplicity, but without restriction of generality, we assume $u$ to be scalar. The total amount of this quantity contained in a given volume $V$ is then

$$M_V(t) \equiv \int_V u(\mathbf{r}, t)\, d\mathbf{r} \tag{5.2}$$

The "flux" through the surface $S$ of the volume is denoted by $J$. It is defined as the net amount entering the volume $V$ per unit time. We further define a "flux density", or "current density" $\mathbf{j}(\mathbf{r}, t)$ as a local contribution to the total influx (see Fig. 5.1):

$$J \equiv -\int_O \mathbf{j}(\mathbf{r}, t) \cdot d\mathbf{S} \quad \text{(per def.)} \tag{5.3}$$

$$= -\int_V (\boldsymbol{\nabla} \cdot \mathbf{j})\, d\mathbf{r} \quad \text{(Gauss law)} \tag{5.4}$$

Restricting the discussion to the particularly important case of an *in toto* conserved quantity, we require the *continuity equation*

$$\frac{dM_V}{dt} = J \tag{5.5}$$

to hold, which is equivalent to

$$\int_V \left[ \frac{\partial u}{\partial t} + \boldsymbol{\nabla} \cdot \mathbf{j} \right] d\mathbf{r} = 0 \tag{5.6}$$

or, since the volume $V$ is arbitrary,

$$\boxed{ \frac{\partial u}{\partial t} = -\boldsymbol{\nabla} \cdot \mathbf{j} } \tag{5.7}$$

We denote equ. 5.7 as the general *conservative PDE*.

In most physically relevant cases the flux density $\mathbf{j}$ will not depend explicitly on $\mathbf{r}$ and $t$, but only implicitly by way of the density $u(\mathbf{r}, t)$ or its spatial derivative, $\boldsymbol{\nabla} u(\mathbf{r}, t)$:

$$\mathbf{j} = \mathbf{j}(u) \quad \text{or} \quad \mathbf{j} = \mathbf{j}(\boldsymbol{\nabla} u) \tag{5.8}$$

In the first instance, $\mathbf{j} = \mathbf{j}(u)$, we are dealing with the *conservative-hyperbolic* equation

$$\boxed{ \frac{\partial u}{\partial t} = -\boldsymbol{\nabla} \cdot \mathbf{j}(u) } \tag{5.9}$$

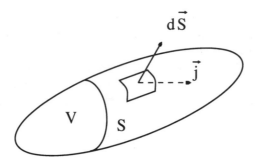

Figure 5.1: Derivation of the conservative PDE

Why "hyperbolic"? At first sight, equation 5.9 does not resemble the standard form as presented by, say, the wave equation

$$\frac{\partial^2 u}{\partial t^2} = c^2 \frac{\partial^2 u}{\partial x^2} \tag{5.10}$$

We can see the connection if we introduce in equ. 5.10 the new variables $r \equiv \partial u/\partial t$ and $s \equiv c(\partial u/\partial x)$. We find

$$\frac{\partial r}{\partial t} = c \frac{\partial s}{\partial x} \tag{5.11}$$

$$\frac{\partial s}{\partial t} = c \frac{\partial r}{\partial x} \tag{5.12}$$

or

$$\frac{\partial \mathbf{u}}{\partial t} = -\frac{\partial \mathbf{j}}{\partial x} \equiv -\mathbf{C} \cdot \frac{\partial \mathbf{u}}{\partial x} \tag{5.13}$$

where

$$\mathbf{u} \equiv \begin{pmatrix} r \\ s \end{pmatrix} , \ \mathbf{j} \equiv \begin{pmatrix} 0 & -c \\ -c & 0 \end{pmatrix} \cdot \mathbf{u} \equiv \mathbf{C} \cdot \mathbf{u} \tag{5.14}$$

EXAMPLE: The equation of motion for a plane electromagnetic wave may be written in two different ways. Denoting by $E_y$ and $B_z$ the non-vanishing components of the electric and magnetic fields, respectively, Maxwell's equations lead to

$$\frac{\partial E_y}{\partial t} = c \frac{\partial B_z}{\partial x} \tag{5.15}$$

$$\frac{\partial B_z}{\partial t} = c \frac{\partial E_y}{\partial x} \tag{5.16}$$

The equations have indeed the form 5.13. However, differentiating by $t$ and $x$ and subtracting one may easily derive the wave equation

$$\frac{\partial^2 E_y}{\partial t^2} = c^2 \frac{\partial^2 E_y}{\partial x^2} \tag{5.17}$$

| hyperbolic | parabolic | elliptic |
|---|---|---|
| conservative-hyperbolic | conservative-parabolic | |
| advective | diffusive | |

Table 5.2: Partial differential equations in physics

As evident from equ. 5.14, the vector $\mathbf{j}(\mathbf{u})$ is a *linear* function of $\mathbf{u}$. Equations with this property are again an important subclass of the conservative-hyperbolic PDEs. They are known as *advective* equations. The numerical schemes to be described in the following sections are applicable to the entire class of conservative-hyperbolic PDEs, but the analysis of stability is most easily demonstrated in the context of advective equations.

An heuristic overview on the various types of PDEs that are of importance in physics is presented in Table 5.2.

## 5.1    Initial Value Problems I: Conservative-hyperbolic DE

We seek to construct algorithms for the general equation

$$\frac{\partial \mathbf{u}}{\partial t} = -\frac{\partial \mathbf{j}}{\partial x} \tag{5.18}$$

or the more specific *advective* equation

$$\frac{\partial \mathbf{u}}{\partial t} = -\mathbf{C} \cdot \frac{\partial \mathbf{u}}{\partial x} \tag{5.19}$$

It will turn out that the "best" (i.e. most stable, exact, etc.) method is the *Lax-Wendroff* technique. To introduce this method it is best to proceed, in ascending order of sophistication, via the *FTCS* scheme, the *Lax* and the *leapfrog* methods.

### 5.1.1    FTCS Scheme; Stability Analysis

Using the notation $\mathbf{u}_j^n \equiv \mathbf{u}(x_j, t_n)$ we may rewrite equ. 5.18 to lowest order as

$$\frac{1}{\Delta t} \left[ \mathbf{u}_j^{n+1} - \mathbf{u}_j^n \right] \approx -\frac{1}{2\,\Delta x} \left[ \mathbf{j}_{j+1}^n - \mathbf{j}_{j-1}^n \right] \tag{5.20}$$

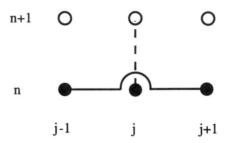

Figure 5.2: FTCS scheme for the conservative-hyperbolic equation

The time derivative is here replaced by $\Delta_n \mathbf{u}^n / \Delta t$ (which explains part of the name: FT for "forward-time"), and in place of $\partial \mathbf{j}/\partial x$ the centered DST approximation $\mu \delta_j \mathbf{j}_j / \Delta x$ is used (CS for "centered-space"). The result of all this is an explicit formula for $\mathbf{u}_j^{n+1}$,

$$\mathbf{u}_j^{n+1} = \mathbf{u}_j^n - \frac{\Delta t}{2\Delta x} \left[ \mathbf{j}_{j+1}^n - \mathbf{j}_{j-1}^n \right] \tag{5.21}$$

which is depicted, in a self-explaining manner, in Figure 5.2.

What about the stability of such a method? The following procedure, due to von Neumann, permits an appropriate generalization of the stability analysis we have used in the context of ordinary differential equations.

Assume, for simplicity, that the solution function $u$ be scalar. At some time $t_n$ the function $u(x,t)$ may be expanded in spatial Fourier components:

$$u_j^n = \sum_k U_k^n e^{ikx_j} \tag{5.22}$$

where $k = 2\pi l/L$ $(l = 0, 1, \ldots)$ is a discrete wave number (see Appendix B, with a slightly different notation). The coefficients $U_k^n$ thus determine the shape of the "snapshot" of $u(x)$ at time $t_n$. If we can obtain, by inserting the Fourier series in the transformation law $u_j^{n+1} = T[u_{j'}^n]$, an according transformation rule for the Fourier components,

$$U_k^{n+1} = g(k)\, U_k^n \tag{5.23}$$

then the stability condition reads

$$|g(k)| \le 1 \ \text{ for all k} \tag{5.24}$$

Applying this idea to the FTCS formula for the advective equation with flux density $j = c\,u$ we find

$$g(k)\, U_k^n e^{ikj\,\Delta x} = U_k^n e^{ikj\,\Delta x} - \frac{c\,\Delta t}{2\,\Delta x} U_k^n \left[ e^{ik(j+1)\Delta x} - e^{ik(j-1)\Delta x} \right] \tag{5.25}$$

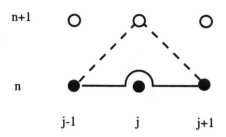

Figure 5.3: Lax scheme (Conservative-hyperbolic equation)

or

$$g(k) = 1 - \frac{ic\,\Delta t}{\Delta x}\,\sin\,k\Delta x \tag{5.26}$$

Obviously, $|g(k)| > 1$ for any $k$; the FTCS method is inherently unstable. Recalling our earlier experiences with another explicit first order method, the Euler-Cauchy scheme of Section 4.1.3, we cannot expect anything better.

## 5.1.2   Lax Scheme

Replacing in the FTCS formula the term $\mathbf{u}_j^n$ by its spatial average $[\mathbf{u}_{j+1}^n + \mathbf{u}_{j-1}^n]/2$, we obtain

$$\mathbf{u}_j^{n+1} = \frac{1}{2}\left[\mathbf{u}_{j+1}^n + \mathbf{u}_{j-1}^n\right] - \frac{\Delta t}{2\Delta x}\left[\mathbf{j}_{j+1}^n - \mathbf{j}_{j-1}^n\right] \tag{5.27}$$

(see Fig. 5.3). The same kind of stability analysis as before (assuming scalar $u$ and $j$, with the advective relation $j = cu$) leads to

$$g(k) = \frac{1}{2}\left[e^{ik\Delta x} + e^{-ik\Delta x}\right] - i\frac{c\Delta t}{\Delta x}\frac{e^{ik\Delta x} - e^{-ik\Delta x}}{2i} \tag{5.28}$$

or

$$g(k) = \cos\,k\Delta x - i\frac{c\Delta t}{\Delta x}\,\sin\,k\Delta x \tag{5.29}$$

The condition $|g(k)| \leq 1$ is tantamount to

$$\frac{|c|\Delta t}{\Delta x} \leq 1 \tag{5.30}$$

This inequality, which will pop up again and again in the stability analysis of integration schemes for the advective equation, is called Courant-Friedrichs-Löwy condition. Its meaning may be appreciated from Figure 5.4. The region below the dashed line encompasses, at time $t_n$, that spatial range which according to $x(t_{n+1}) = x(t_n) \pm |c|\,\Delta t$ may in principle contribute to the value of the solution

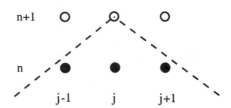

Figure 5.4: Courant-Friedrichs-Löwy condition

function $u_j^{n+1}$ at the next time step. For a large propagation speed $c$ this region is, of course, larger than for small $c$. If a numerical algorithm fails to take into account *all* values $u_{j'}^n$ situated within the relevant region, it will be unstable.

Comparing the Lax scheme to the FTCS formula, we find an apparently spurious term which cannot be accounted for by considering the original DE:

$$\frac{u_j^{n+1} - u_j^n}{\Delta t} = -c \frac{u_{j+1}^n - u_{j-1}^n}{2\Delta x} + \frac{1}{2} \frac{u_{j+1}^n - 2u_j^n + u_{j-1}^n}{\Delta t} \tag{5.31}$$

The second fraction on the right-hand side has the form of a diffusion term,

$$\frac{\delta_j^2 u_j^n}{2\Delta t} \approx \frac{\partial^2 u}{\partial x^2} \frac{(\Delta x)^2}{2\Delta t} \tag{5.32}$$

implying that by using the Lax method we are in fact solving the equation

$$\frac{\partial u}{\partial t} = -c \frac{\partial u}{\partial x} + \frac{(\Delta x)^2}{2\Delta t} \frac{\partial^2 u}{\partial x^2} \tag{5.33}$$

However, for small enough $(\Delta x)^2/\Delta t$ this additional term – which obviously brought us the gift of stability – will be negligible. We require therefore that in addition to the stability condition $|c|\Delta t \le \Delta x$ we have

$$|c|\Delta t >> \frac{\Delta x}{2} \frac{|\delta^2 u|}{|\delta u|} \tag{5.34}$$

Incidentally, the Lax scheme amplification factor $g(k)$ for small $k$, i.e. for long wave length modes, is always near to 1:

$$g(k) \approx 1 - \frac{(k\Delta x)^2}{2} - i \frac{c\Delta t}{\Delta x} k\Delta x \approx 1 \tag{5.35}$$

This means that aberrations from the correct solution that range over many grid points will die off very slowly. This flaw can be mended only by introducing algorithms of higher order (see below).

EXAMPLE: The one-dimensional wave equation may be written in advective form as

$$\frac{\partial \mathbf{u}}{\partial t} = -\mathbf{C} \cdot \frac{\partial \mathbf{u}}{\partial x}$$

where

$$\mathbf{u} = \begin{pmatrix} r \\ s \end{pmatrix} \quad and \quad \mathbf{C} = \begin{pmatrix} 0 & -c \\ -c & 0 \end{pmatrix}$$

(see 5.13-5.14). The Lax scheme for this equation reads

$$r_j^{n+1} = \frac{1}{2}[r_{j+1}^n + r_{j-1}^n] + \frac{c\Delta t}{2\Delta x}[s_{j+1}^n - s_{j-1}^n] \tag{5.36}$$

$$s_j^{n+1} = \frac{1}{2}[s_{j+1}^n + s_{j-1}^n] + \frac{c\Delta t}{2\Delta x}[r_{j+1}^n - r_{j-1}^n] \tag{5.37}$$

To make the connection to the above-mentioned example of a plane electromagnetic wave we may interpret $r$ and $s$ as the magnetic and electric field strengths, respectively, and $c$ as the speed of light. Of course, this particular equation may be solved easily by analytic methods; but any slight complication, such as a locally varying light velocity, will render the *numerical* procedure all but irresistible.

   *Warning:* Since the Lax scheme is sensitive to the ratio $|\delta^2 u|/|\delta u|$, it may fail for quite simple wave propagation problems.

## 5.1.3    Leapfrog Scheme (LF)

Both in the FTCS and in the Lax scheme a first order approximation was used for the time derivative: $\partial u/\partial t \approx \Delta_n u_j^n / \Delta t$. Remembering the excellent record of the second-order Stirling formula (see Sec. 1.2.3) we insert

$$\frac{\partial u}{\partial t} \approx \frac{u^{n+1} - u^{n-1}}{2\Delta t} \tag{5.38}$$

in the above equation, to find the *leapfrog* expression

$$\boxed{\mathbf{u}_j^{n+1} - \mathbf{u}_j^{n-1} = -\frac{\Delta t}{\Delta x}\left[\mathbf{j}_{j+1}^n - \mathbf{j}_{j-1}^n\right]} \tag{5.39}$$

(Similar formulae were developed earlier for ordinary DE; see equ. 4.26 and Fig. 4.5.)

   The amplification factor $g(k)$ obeys (assuming $j = cu$)

$$g^2 - 1 = g\frac{c\Delta t}{\Delta x}(e^{ik\Delta x} - e^{-ik\Delta x}) \tag{5.40}$$

or, with $a \equiv c\Delta t/\Delta x$,

$$g(k) = -ia \sin k\Delta x \pm \sqrt{1 - (a \sin k\Delta x)^2} \tag{5.41}$$

The requirement $|g|^2 \leq 1$ results once more in the CFL condition,

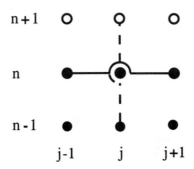

Figure 5.5: Leapfrog scheme for the conservative-hyperbolic equation

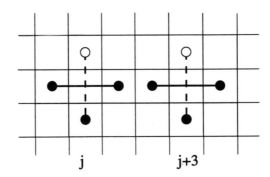

Figure 5.6: Decoupled space-time grids in the leapfrog scheme

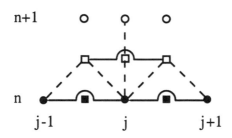

Figure 5.7: Lax-Wendroff scheme

$$\frac{c\,\Delta t}{\Delta x} \leq 1 \tag{5.42}$$

One drawback of the leapfrog technique is that it describes the evolution of the solution function on two decoupled space-time grids (see Fig. 5.6). The solutions on the "black" and "white" fields will increasingly differ in the course of many time steps. An ad hoc remedy is to discard one of the two solutions, giving up half the information attainable with the given grid finesse. A better way is to connect the two subgrids by adding a weak coupling term, which once more has the form of a diffusion contribution, on the right-hand side of 5.39:

$$\ldots + \varepsilon[\mathbf{u}_{j+1}^n - 2\mathbf{u}_j^n + \mathbf{u}_{j-1}^n] \tag{5.43}$$

## 5.1.4   Lax-Wendroff Scheme (LW)

A somewhat more complex second-order procedure which, however, avoids the disadvantages of the methods described so far, is explained in the Figures 5.7 and 5.8.

Stability analysis is now a bit more involved than for the previous techniques. Assuming once more that $j = cu$ and using the ansatz $U_k^{n+1} = g(k)U_k^n$ one inserts the Fourier series 5.22 in the successive stages of the LW procedure. This yields

$$g(k) = 1 - ia \sin k\Delta x - a^2(1 - \cos k\Delta x), \tag{5.47}$$

with $a = c\Delta t/\Delta x$. The requirement $|g|^2 \leq 1$ leads once again to the CFL condition 5.30.

## 5.1.5   Lax and Lax-Wendroff in Two Dimensions

For simplicity we will again assume a scalar solution $u(\mathbf{r}, t)$. The conservative-hyperbolic equation reads, in two dimensions,

$$\frac{\partial u}{\partial t} = -\frac{\partial j_x}{\partial x} - \frac{\partial j_y}{\partial y} \tag{5.48}$$

**Lax-Wendroff scheme:**

- Lax method with half-step: $\Delta x/2$, $\Delta t/2$:

$$\mathbf{u}_{j+1/2}^{n+1/2} = \frac{1}{2}\left[\mathbf{u}_{j+1}^n + \mathbf{u}_j^n\right] - \frac{\Delta t}{2\Delta x}\left[\mathbf{j}_{j+1}^n - \mathbf{j}_j^n\right] \qquad (5.44)$$

  and analogously for $\mathbf{u}_{j-1/2}^{n+1/2}$ .

- Evaluation, e.g. for the advective case $\mathbf{j} = \mathbf{C} \cdot \mathbf{u}$:

$$\mathbf{u}_{j+1/2}^{n+1/2} \Longrightarrow \mathbf{j}_{j+1/2}^{n+1/2} \qquad (5.45)$$

- *leapfrog* with half-step:

$$\mathbf{u}_j^{n+1} = \mathbf{u}_j^n - \frac{\Delta t}{\Delta x}\left[\mathbf{j}_{j+1/2}^{n+1/2} - \mathbf{j}_{j-1/2}^{n+1/2}\right] \qquad (5.46)$$

Figure 5.8: Lax-Wendroff method

(where in the advective case $j_x = c_x u$ and $j_y = c_y u$.) The Lax scheme is now written as

$$u_{i,j}^{n+1} = \frac{1}{4}\left[u_{i+1,j}^n + u_{i,j+1}^n + u_{i-1,j}^n + u_{i,j-1}^n\right] - \frac{\Delta t}{2\Delta x}\left[j_{x,i+1,j}^n - j_{x,i-1,j}^n\right]$$
$$- \frac{\Delta t}{2\Delta y}\left[j_{y,i,j+1}^n - j_{y,i,j-1}^n\right] \qquad (5.49)$$

In the more efficient Lax-Wendroff algorithm we require, as input for the second stage (half-step leapfrog), quantities such as $j_{x,i+1/2,j-1/2}^{n+1/2}$. These would have to be computed, via $u_{i+1/2,j-1/2}^{n+1/2}$, from $u_{i,j-1/2}^n$, $u_{i+1,j-1/2}^n$ etc. Here we have a problem: quantities with half-step spatial indices ($i+1/2$, $j-1/2$ etc.) are given at half-step times ($t_{n+1/2}$) only. To mend this, one modifies the Lax-Wendroff prescription according to Figs. 5.10-5.11. To calculate $u_{i,j}^{n+1}$, only the points ○ (at $t_n$) are used, while $u_{i+1,j}^{n+1}$ is computed using the points □. This again results in a slight drifting apart of the subgrids ○ and □. If the given differential equation happens to contain a diffusive term, the two grids are automatically coupled. If there is no diffusive contribution, it may be invented, as in the leapfrog method [POTTER 80].

Stability analysis proceeds in the same way as in the one-dimensional case, except for the Fourier modes being now 2-dimensional:

$$u(x,y) = \sum_k \sum_l U_{k,l} e^{ikx+ily} \qquad (5.53)$$

Further analysis results in a suitably generalized CFL condition [POTTER 80],

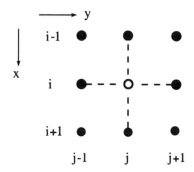

Figure 5.9: Lax method in two dimensions

---

**Lax-Wendroff in 2 dimensions:**

- Lax method to determine the $u$-values at half-step time $t_{n+1/2}$:

$$
\begin{aligned}
u_{i+1,j}^{n+1/2} = {} & \frac{1}{4}\left[u_{i+2,j}^n + u_{i+1,j+1}^n + u_{i,j}^n + u_{i+1,j-1}^n\right] \\
& - \frac{\Delta t}{2\Delta x}\left[j_{x,i+2,j}^n - j_{x,i,j}^n\right] \\
& - \frac{\Delta t}{2\Delta y}\left[j_{y,i+1,j+1}^n - j_{y,i+1,j-1}^n\right]
\end{aligned}
\tag{5.50}
$$

  etc.

- Evaluation at half-step time:

$$
u_{i+1,j}^{n+1/2}, \ldots \implies j_{x,i+1,j}^{n+1/2}, \ldots
\tag{5.51}
$$

- *leapfrog* with half-step:

$$
\begin{aligned}
u_{i,j}^{n+1} = u_{i,j}^n & - \frac{\Delta t}{2\Delta x}\left[j_{x,i+1,j}^{n+1/2} - j_{x,i-1,j}^{n+1/2}\right] \\
& - \frac{\Delta t}{2\Delta y}\left[j_{y,i,j+1}^{n+1/2} - j_{y,i,j-1}^{n+1/2}\right]
\end{aligned}
\tag{5.52}
$$

Figure 5.10: Lax-Wendroff in two dimensions

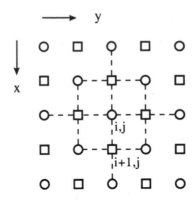

Figure 5.11: First stage (= Lax) in the 2-dimensional LW method: $\bigcirc$... $t_n, t_{n+1}$, $\square$...$t_{n+1/2}$

namely (assuming $\Delta x = \Delta y$)

$$\Delta t \leq \frac{\Delta x}{\sqrt{2}\,\sqrt{c_x^2 + c_y^2}} \tag{5.54}$$

## 5.2  Initial Value Problems II: Conservative-parabolic DE

The generic equation of this kind is the *diffusive* equation

$$\frac{\partial u}{\partial t} = \frac{\partial}{\partial x}\left(\lambda \frac{\partial u}{\partial x}\right) \tag{5.55}$$

which for a constant transport coefficient $\lambda$ assumes the even simpler form

$$\frac{\partial u}{\partial t} = \lambda \frac{\partial^2 u}{\partial x^2} \tag{5.56}$$

In the case of parabolic equations there are more feasible integration algorithms to choose from than there were for hyperbolic equations. The method that may be regarded "best" in more than one respect is the second-order algorithm by Crank and Nicholson. However, there is also another quite competitive method of second order, called Dufort-Frankel scheme, and even the various first-order methods, which for didactic reasons will be treated first, are reasonably stable.

### 5.2.1  FTCS Scheme

We can once more derive a "forward time-centered space" algorithm, replacing $\partial u/\partial t$ by the DNGF approximation $\Delta_n u/\Delta t$, and $\partial^2 u/\partial x^2$ by the DDST formula

Figure 5.12: FTCS method for the parabolic-diffusive equation

$\delta_j^2 u/(\Delta x)^2$:

$$\frac{1}{\Delta t}\left[u_j^{n+1} - u_j^n\right] = \frac{\lambda}{(\Delta x)^2}\left[u_{j+1}^n - 2u_j^n + u_{j-1}^n\right] \tag{5.57}$$

Using $a \equiv \lambda \Delta t/(\Delta x)^2$ this may be written as

$$u_j^{n+1} = (1 - 2a)u_j^n + a(u_{j-1}^n + u_{j+1}^n) \tag{5.58}$$

(see Fig. 5.12). In contrast to the hyperbolic case the FTCS method is *stable* for parabolic-diffusive equations. For the $k$-dependent growth factor we find

$$g(k) = 1 - 4a \, \sin^2 \frac{k\Delta x}{2} \tag{5.59}$$

which tells us that for stability the condition

$$2\lambda \frac{\Delta t}{(\Delta x)^2} \leq 1 \tag{5.60}$$

must be met. Noting that the characteristic time for the diffusion over a distance $\Delta x$ (i.e. one lattice space) is

$$\tau = \frac{(\Delta x)^2}{2\lambda} \tag{5.61}$$

we understand that $\Delta t \leq \tau$ is required for stability.

If we try to enhance the spatial resolution by reducing $\Delta x$, the characteristic time will decrease quadratically, leading to an unpleasant reduction of the permitted time step. The FTCS scheme is therefore, though simple and stable, rather inefficient.

To allow for an explicit or implicit spatial variation of $\lambda$ we may write the FTCS formula as

$$u_j^{n+1} = u_j^n + \frac{\Delta t}{(\Delta x)^2}\left[\lambda_{j+1/2}(u_{j+1}^n - u_j^n) - \lambda_{j-1/2}(u_j^n - u_{j-1}^n)\right] \tag{5.62}$$

where

$$\lambda_{j+1/2} \equiv \lambda(x_{j+1/2}) \quad \text{or} \quad \lambda_{j+1/2} \equiv \lambda(u_{j+1/2}) \tag{5.63}$$

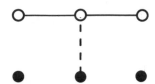

Figure 5.13: Implicit method for the parabolic-diffusive equation

denotes a suitably interpolated interlattice value of $\lambda$.

EXERCISE: Apply the FTCS scheme to the thermal conduction problem of Sec. 1.4.2. Interpret the behavior of the solution for varying time step sizes in the light of the above stability considerations.

## 5.2.2    Implicit Scheme of First Order

We obtain a considerable increase in efficiency if we take the second spatial derivative at time $t_{n+1}$ instead of $t_n$:

$$\frac{1}{\Delta t}\left[u_j^{n+1} - u_j^n\right] = \frac{\lambda}{(\Delta x)^2}\left[u_{j+1}^{n+1} - 2u_j^{n+1} + u_{j-1}^{n+1}\right] \tag{5.64}$$

(see Fig. 5.13). Again defining $a \equiv \lambda \Delta t/(\Delta x)^2$, we find, for each space point $x_j$ $(j = 1, 2, ..N - 1)$,

$$-au_{j-1}^{n+1} + (1 + 2a)u_j^{n+1} - au_{j+1}^{n+1} = u_j^n \tag{5.65}$$

Let the boundary values $u_0$ and $u_N$ be given; the set of equations may then be written as

$$\mathbf{A} \cdot \mathbf{u}^{n+1} = \mathbf{u}^n \tag{5.66}$$

with

$$\mathbf{A} \equiv \begin{pmatrix} 1 & 0 & 0 & \cdot & \cdot & 0 \\ -a & 1+2a & -a & 0 & \cdot & 0 \\ 0 & \cdot & \cdot & \cdot & 0 & \cdot \\ \cdot & \cdot & \cdot & \cdot & \cdot & \cdot \\ \cdot & \cdot & \cdot & -a & 1+2a & -a \\ \cdot & \cdot & \cdot & 0 & 0 & 1 \end{pmatrix} \tag{5.67}$$

We have seen before that a tridiagonal system of this kind is most easily inverted by *recursion* (see Section 2.1.4).

Asking for error propagation, we find

$$-a\,g\,e^{-ik\Delta x} + (1 + 2a)g - a\,g\,e^{ik\Delta x} = 1 \tag{5.68}$$

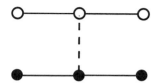

Figure 5.14: Crank-Nicholson technique for the parabolic-diffusive equation

or

$$g = \frac{1}{1 + 4a\,\sin^2(k\Delta x/2)} \tag{5.69}$$

Since $|g| \leq 1$ under all circumstances, we have here an unconditionally stable algorithm!

Interestingly, the method retains its consistency regarding the limit $\Delta x \to 0$ even if we make the time step $\Delta t$ very large. In that case

$$u_{j+1}^{n+1} - 2u_j^{n+1} + u_{j-1}^{n+1} \equiv \delta_j^2 u_j^{n+1} = 0 \tag{5.70}$$

which corresponds neatly to the differential equation $\partial^2 u/\partial x^2 = 0$ describing the long time (stationary) behavior of the diffusion equation.

EXERCISE: Apply the implicit technique to the thermal conduction problem discussed in Sects. 5.2.1 and 1.4.2. Consider the efficiency of the procedure as compared to FTCS. Relate the problem to the *random walk* of p. 77.

## 5.2.3    Crank-Nicholson Scheme (CN)

As before, we replace $\partial u/\partial t$ by $\Delta_n u/\Delta t \equiv (u^{n+1} - u^n)/\Delta t$. However, noting that this approximation is in fact centered at $t_{n+1/2}$, we introduce the same kind of time centering on the right-hand side of 5.56. Taking the mean of $\delta_j^2 u^n$ (= FTCS) and $\delta_j^2 u^{n+1}$ (= implicit scheme) we write

$$\frac{1}{\Delta t}\left[u_j^{n+1} - u_j^n\right] = \frac{\lambda}{2(\Delta x)^2}\left[(u_{j+1}^{n+1} - 2u_j^{n+1} + u_{j-1}^{n+1}) + (u_{j+1}^n - 2u_j^n + u_{j-1}^n)\right] \tag{5.71}$$

(see Fig. 5.14). A closer look reveals that this *Crank-Nicholson* formula is now of *second* order in $\Delta t$ [PRESS 86]. Defining $a \equiv \lambda\Delta t/2(\Delta x)^2$ (note the factor $1/2$ as compared to earlier definitions!) we may write the CN algorithm as

$$\boxed{-au_{j-1}^{n+1} + (1+2a)u_j^{n+1} - au_{j+1}^{n+1} = au_{j-1}^n + (1-2a)u_j^n + au_{j+1}^n} \tag{5.72}$$

In matrix notation this is

$$\mathbf{A}\cdot\mathbf{u}^{n+1} = \mathbf{B}\cdot\mathbf{u}^n \tag{5.73}$$

with

$$
A \equiv \begin{pmatrix}
1 & 0 & 0 & . & . & 0 \\
-a & 1+2a & -a & 0 & . & 0 \\
0 & . & . & . & 0 & . \\
. & . & . & . & . & . \\
. & . & . & -a & 1+2a & -a \\
. & . & . & 0 & 0 & 1
\end{pmatrix}
\quad
B \equiv \begin{pmatrix}
1 & 0 & 0 & . & . & 0 \\
a & 1-2a & a & 0 & . & 0 \\
0 & . & . & . & 0 & . \\
. & . & . & . & . & . \\
. & . & . & a & 1-2a & a \\
. & . & . & 0 & 0 & 1
\end{pmatrix}
$$

Thus we have to solve, at each time step, a tridiagonal system of equations. The *recursion technique* of Section 2.1.4 does the trick fast enough.

Whenever the transport coefficient $\lambda$ depends – either explicitly or implicitly via $u$ – on position, the CN algorithm may be adapted accordingly [PRESS 86].

The amplification factor is

$$
g(k) = \frac{1 - 2a \sin^2(k\Delta x/2)}{1 + 2a \sin^2(k\Delta x/2)} \leq 1 ,
\tag{5.74}
$$

which makes the CN method unconditionally stable.

For large time steps the CN algorithm is not quite as well-behaved as the first-order implicit scheme. $\Delta t \to \infty$ results in

$$
-\delta_j^2 u_j^{n+1} = \delta_j^2 u_j^n
\tag{5.75}
$$

yielding

$$
\lim_{\Delta t \to \infty} |g(k)| = 1
\tag{5.76}
$$

In this limit the method is only marginally stable – errors do not grow, but do not decay either.

EXAMPLE: The time-dependent Schroedinger equation,

$$
\frac{\partial u}{\partial t} = -iHu , \quad \text{with } H \equiv \frac{\partial^2}{\partial x^2} + U(x)
\tag{5.77}
$$

when rewritten à la Crank-Nicholson, reads

$$
\begin{aligned}
\frac{1}{\Delta t}[u_j^{n+1} - u_j^n] &= -\frac{i}{2}[(Hu)_j^{n+1} + (Hu)_j^n] \\
&= -\frac{i}{2}\left[ \frac{\delta_j^2 u_j^{n+1}}{(\Delta x)^2} + U_j u_j^{n+1} + \frac{\delta_j^2 u_j^n}{(\Delta x)^2} + U_j u_j^n \right]
\end{aligned}
\tag{5.78}
$$

With $a \equiv \Delta t/2(\Delta x)^2$ and $b_j \equiv U(x_j)\Delta t/2$ this leads to

$$
\begin{aligned}
(ia)u_{j-1}^{n+1} + (1 - 2ia + ib_j)u_j^{n+1} + (ia)u_{j+1}^{n+1} = \\
= (-ia)u_{j-1}^n + (1 + 2ia - ib_j)u_j^n + (-ia)u_{j+1}^n
\end{aligned}
\tag{5.79}
$$

Again, we have a tridiagonal system which may be inverted very efficiently by the recursion method of Sec. 2.1.4.

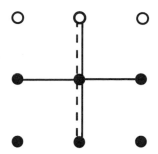

Figure 5.15: Dufort-Frankel technique for the parabolic-diffusive equation

## 5.2.4    Dufort-Frankel Scheme (DF)

The DF scheme is similar to the *leapfrog* algorithm – which, however, would be unstable when applied without precaution to the diffusive equation. We write

$$\frac{1}{2\Delta t}\left[u_j^{n+1} - u_j^{n-1}\right] = \frac{\lambda}{(\Delta x)^2}\left[u_{j+1}^n - (u_j^{n+1} + u_j^{n-1}) + u_{j-1}^n\right] \qquad (5.80)$$

Note that instead of the term $-2u_j^n$ we have introduced the combination $-(u_j^{n+1} + u_j^{n-1})$ (see Fig. 5.15). Using $a \equiv 2\lambda\Delta t/(\Delta x)^2$ this may be written as

$$u_j^{n+1} = \frac{1-a}{1+a}u_j^{n-1} + \frac{a}{1+a}\left[u_{j+1}^n + u_{j-1}^n\right] \qquad (5.81)$$

The DF algorithm is of second order in $\Delta t$, just as the CN scheme. It has the advantage over CN that 5.81 is an *explicit* expression for $u_j^{n+1}$ – albeit with the necessity to store the past values $u_j^{n-1}$.

The amplification factor is

$$g = \frac{1}{1+a}\left[a\cos k\Delta x \pm \sqrt{1 - a^2 \sin^2 k\Delta x}\right] \qquad (5.82)$$

Considering in turn the cases $a^2 \sin^2 k\Delta x \geq 1$ and $\ldots < 1$ we find that $|g|^2 \leq 1$ always; the method is unconditionally stable.

## 5.3    Boundary Value Problems: Elliptic DE

The standard problem we will consider to demonstrate the various methods for elliptic equations is the two-dimensional potential equation,

$$\frac{\partial^2 u}{\partial x^2} + \frac{\partial^2 u}{\partial y^2} = -\rho(x, y) \qquad (5.83)$$

For finite charge densities $\rho(x, y)$ this is Poisson's equation; in charge-free space $\rho \equiv 0$ it is called Laplace's equation.

Written in terms of finite differences (assuming $\Delta y = \Delta x \equiv \Delta l$) equ. 5.83 reads

$$\frac{1}{(\Delta l)^2} \left[ \delta_i^2 u_{i,j} + \delta_j^2 u_{i,j} \right] = -\rho_{i,j} \tag{5.84}$$

or

$$\frac{1}{(\Delta l)^2} \left[ u_{i+1,j} - 2u_{i,j} + u_{i-1,j} + u_{i,j+1} - 2u_{i,j} + u_{i,j-1} \right] = -\rho_{i,j} \tag{5.85}$$

$$(i = 1, 2, \ldots N; \quad j = 1, 2, \ldots M)$$

In enumerating the lattice points one may apply the rules familiar from matrices, such that the coordinate $y$ and the index $j$ increase to the right, and $x$ and $i$ downwards.

We now construct a vector $\mathbf{v}$ of length $N.M$ by linking together the *rows* of the matrix $\{u_{i,j}\}$:

$$v_r = u_{i,j}, \quad \text{with } r = (i - 1)M + j \tag{5.86}$$

Conversely,

$$i = \text{int}\left(\frac{r-1}{M}\right) + 1 \text{ and } j = [(r-1) \bmod M] + 1 \tag{5.87}$$

where int(...) denotes the next smaller integer. Equation 5.85 then transforms to

$$v_{r-M} + v_{r-1} - 4v_r + v_{r+1} + v_{r+M} = -(\Delta l)^2 \rho_r \tag{5.88}$$

which may be written

$$\mathbf{A} \cdot \mathbf{v} = \mathbf{b} \tag{5.89}$$

with the vector $\mathbf{b} \equiv -(\Delta l)^2 \{\rho_1, \ldots \rho_{N.M}\}^T$ and the pentadiagonal matrix

$$\mathbf{A} \equiv \begin{pmatrix} -4 & 1 & \cdots & 1 & \\ 1 & -4 & 1 & & \ddots \\ \vdots & \ddots & \ddots & \ddots & \\ 1 & & & & \\ & \ddots & & & \end{pmatrix} \tag{5.90}$$

What about the boundaries of the physical system we are considering? The equations 5.85, which lead to the specific form 5.90 for the matrix $\mathbf{A}$, apply in this form only to the interior region of the lattice. At the rim of the grid – and thus in certain parts of $\mathbf{A}$ – the most fundamental (Dirichlet) boundary conditions will provide us with obligatory values for the solution $u_{i,j}^0$. (In this context the superscript 0 denotes a required value, and not the time $t = 0$). Assume that the grid consists of only $5 \times 5$ points on a square lattice, with $u_{i,j} = u_{i,j}^0$ being given along the sides

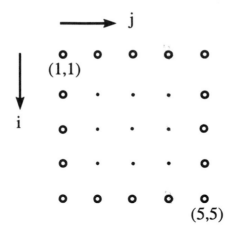

Figure 5.16: Potential equation on a $5 \times 5$ lattice: at the points $\circ$ the values of the potential $u(x, y)$ are given (Dirichlet boundary conditions)

of the square (see Fig. 5.16). This gives us a number of trivial equations of the type $v_1 = u_{1,1}^0$ for the points on the rim. At the interior points equ. 5.89 holds:

$$-4v_7 + v_8 + v_{12} = -(\Delta l)^2 \rho_{2,2} - u_{2,1}^0 - u_{1,2}^0 \qquad (5.91)$$

etc. More specifically, the matrix $\mathbf{A}$ has the form given in Fig. 5.17. The vector $\mathbf{v}$ consists of the nine elements $v_7, v_8, v_9, v_{12}, v_{13}, v_{14}, v_{17}, v_{18}, v_{19}$, and the vector $\mathbf{b}$ has components

$$
\begin{aligned}
b_7 &= -(\Delta l)^2 \rho_7 - u_{1,2}^0 - u_{2,1}^0 \\
b_8 &= -(\Delta l)^2 \rho_8 - u_{1,3}^0 \\
b_9 &= -(\Delta l)^2 \rho_9 - u_{1,4}^0 - u_{2,5}^0 \\
b_{12} &= -(\Delta l)^2 \rho_{12} - u_{3,1}^0 \\
b_{13} &= -(\Delta l)^2 \rho_{13} \\
b_{14} &= -(\Delta l)^2 \rho_{14} - u_{3,5}^0 \\
b_{17} &= -(\Delta l)^2 \rho_{17} - u_{4,1}^0 - u_{5,2}^0 \\
b_{18} &= -(\Delta l)^2 \rho_{18} - u_{5,3}^0 \\
b_{19} &= -(\Delta l)^2 \rho_{19} - u_{4,5}^0 - u_{5,4}^0
\end{aligned}
$$

So far we have considered boundary conditions of the *Dirichlet* type. If we are dealing with *Neumann* boundary conditions of the form

$$\left(\frac{\partial u}{\partial x}\right)_{i,j} = \alpha_{i,j} \; ; \quad \left(\frac{\partial u}{\partial y}\right)_{i,j} = \beta_{i,j} \qquad (5.92)$$

$$
\begin{pmatrix}
-4 & 1 & & | & 1 & & & | & & & \\
1 & -4 & 1 & | & & 1 & & | & & & \\
 & 1 & -4 & | & & & 1 & | & & & \\
- & - & - & | & - & - & - & | & - & - & - \\
1 & & & | & -4 & 1 & & | & 1 & & \\
 & 1 & & | & 1 & -4 & 1 & | & & 1 & \\
 & & 1 & | & & 1 & -4 & | & & & 1 \\
- & - & - & | & - & - & - & | & - & - & - \\
 & & & | & 1 & & & | & -4 & 1 & \\
 & & & | & & 1 & & | & 1 & -4 & 1 \\
 & & & | & & & 1 & | & & 1 & -4 \\
- & - & - & | & - & - & - & | & - & - & - 
\end{pmatrix}
$$

Figure 5.17: Treatment of Dirichlet-type boundary conditions $u_{i,j} = u_{i,j}^0$ in the case of a $5 \times 5$ lattice

a linear approximation for $u(x,y)$ is used to link the boundary values of $u_{i,j}$ to the adjacent interior points. In the context of the previous example, the *derivatives* are now given along the contour of the square. One proceeds as follows:

- The given grid is enlarged by a surrounding layer of additional lattice points. For the function $u$ at these external points, $u_{0,1}, u_{0,2}, \ldots$, one writes

$$
\begin{aligned}
u_{0,1} &= u_{2,1} - 2\alpha_{1,1}\,\Delta l \\
u_{0,2} &= u_{2,2} - 2\alpha_{1,2}\,\Delta l
\end{aligned}
$$

$$\vdots$$

$$
u_{1,0} = u_{1,2} - 2\beta_{1,1}\,\Delta l
$$

$$\vdots$$

- At the original boundary points, such as $(1,1)$, we have

$$
u_{2,1} - 2u_{1,1} + u_{0,1} + u_{1,2} - 2u_{1,1} + u_{1,0} = -\rho_{1,1}(\Delta l)^2 \tag{5.93}
$$

Elimination of the external values yields

$$
u_{2,1} - 2u_{1,1} + u_{2,1} + u_{1,2} - 2u_{1,1} + u_{1,2} = -\rho_{1,1}(\Delta l)^2 + 2\alpha_{1,1}\,\Delta l + 2\beta_{1,1}\,\Delta l \tag{5.94}
$$

Thus the form of the discretized Poisson equation at the boundary points is the same as in the interior region (equ. 5.85), except that on the right-hand side of 5.94 we now have a modified, "effective" charge density. Again introducing the vector $\mathbf{v}$ and the system matrix $\mathbf{A}$, we find that the upper left-hand corner of $\mathbf{A}$ looks as shown in Fig. 5.18.

$$
\left(
\begin{array}{ccccc|ccccc|ccc}
-4 & 2 & & & & 2 & & & & & & & \\
1 & -4 & 1 & & & & 2 & & & & & & \\
 & 1 & -4 & 1 & & & & 2 & & & & & \\
 & & 1 & -4 & 1 & & & & 2 & & & & \\
 & & & 2 & -4 & & & & & 2 & & & \\
\hline
1 & & & & & -4 & 2 & & & & 2 & & \\
 & 1 & & & & 1 & -4 & 1 & & & & 2 & \\
\end{array}
\right)
$$

Figure 5.18: Treatment of Neumann-type boundary conditions in the case of a
5 × 5 lattice

## 5.3.1   Relaxation and Multigrid Techniques

This is the big moment for the relaxation methods of Section 2.2. Having transformed the given physical equation into the set of linear equations 5.89, we can
apply any one of those iterative techniques to solve for the vector $\mathbf{v}$. In particular,
the Jacobi scheme reads

$$
\mathbf{v}^{n+1} = \left[ \mathbf{I} + \frac{1}{4}\mathbf{A} \right] \cdot \mathbf{v}^n + \frac{(\Delta l)^2}{4}\rho \tag{5.95}
$$

Since the matrix $\mathbf{A}$ is sparsely populated, the Gauss-Seidel and the SOR methods
are just as easy to implement.

However, relaxation methods for the potential and similar equations are lame
on one leg. They perform fast for certain types of iterated solution vectors $\mathbf{v}^n$ but
slow for others. *Multigrid techniques* go a long way to mend this shortcoming.

Let us denote by $\mathbf{e} \equiv \mathbf{v} - \mathbf{v}^n$ the residual error after the $n$-th relaxation step.
As the iteration proceeds, this vector should approach zero. In Chapter 2 we
have already discussed the rate of convergence in very general terms, based solely
on the properties of the system matrix $\mathbf{A}$. We learned that the estimated rate
of convergence is linked to the largest (by absolute value) eigenvalue of $\mathbf{A}$ – or
rather, of the iteration matrix constructed from $\mathbf{A}$. The farther this *spectral radius*
deviates from zero, the more slowly will the relaxation converge.

Inspecting the Jacobi and Gauss-Seidel iteration matrices pertaining to the
specific system matrix 5.90, we find that their spectral radii are close to 1, which
spells bad performance. However, close scrutiny shows that this is just a worst-
case estimate. Convergence depends also on the properties of the iterated vector
$\mathbf{e}$, and by applying multigrid schemes we may manipulate these properties so as to
improve convergence by orders of magnitude.

For the moment, let us denote the total number of grid points by $K \equiv N \cdot M$;
then the iterated vectors $\mathbf{v}^n$ and $\mathbf{e}$ are both of dimension $K$. Like any table of
scalars, $\mathbf{e} \equiv (e_0, \ldots e_{K-1})$ may be written as a sum of Fourier components, or

modes,

$$e_j = \frac{1}{K} \sum_{k=0}^{K-1} E_k \exp^{-2\pi ijk/K} \tag{5.96}$$

where

$$E_k = \sum_{j=0}^{K-1} e_j \exp^{2\pi ijk/K} \tag{5.97}$$

(see Appendix B). Further analysis shows that relaxation will be faster for the "oscillatory" high wave number modes of **e** than for the "smooth", low wave number components. In other words, embarking on a relaxation procedure with a certain starting approximation $\mathbf{v}^0$ we will find that any spike- or ripple-like deviations from the exact solution will die off much faster than the worst case estimate would make us expect, while the smooth components of the error will indeed linger on for many further iterations.

This is where multigrid methods come in. The idea is that "smoothness" is really just a matter of scale. Rewriting the original problem such that the lattice width is now double the old one, each wave number is automatically increased by a factor of two; this may be verified by replacing $K$ by $K/2$ in 5.97. The respective modes are therefore processed more efficiently on the coarser grid than in the original representation. The trick, then, is to switch back and forth between representations of the physical property **v** on grids of different spacings. First the oscillatory modes are taken care of by several iterations on the fine grid, until only the smooth modes remain. Then the grid spacing is doubled – by considering only every other lattice point, or by a more refined recipe – and the remaining long-wavelength modes are relaxed through a number of steps. A cascade of several such coarsening stages may be passed before the fine grid is gradually reconstructed using some interpolation scheme.

There is a rich literature on various implementations of the multigrid technique. A short introductory overview is [BRIGGS 94]; more technical details may be found in [WESSELING 92].

## 5.3.2    ADI Method for the Potential Equation

We are now ready to keep the promise made in Section 2.2.4, to demonstrate the use of the particularly effective *alternating direction implicit* technique in the context of the potential equation.

In addition to the previously defined vector **v** we construct another long vector **w** by linking together the *columns* of the matrix $\{u_{i,j}\}$:

$$w_s = u_{i,j}, \quad \text{with } s = (j-1)N + i \tag{5.98}$$

and conversely

$$j = \text{int}\left(\frac{s-1}{N}\right) + 1; \quad i = [(s-1) \bmod N] + 1 \tag{5.99}$$

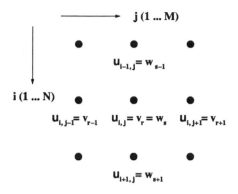

Figure 5.19: ADI method

The vectors $\mathbf{v}$ and $\mathbf{w}$ have equal status. They are related to each other by the *reordering transformation*

$$\mathbf{w} = \mathbf{U} \cdot \mathbf{v} \qquad (5.100)$$

where $\mathbf{U}$ is a sparse matrix consisting solely of elements 0 and 1.
    With this the discretized potential equation 5.85 may be written as

$$w_{s+1} - 2w_s + w_{s-1} + v_{r+1} - 2v_r + v_{r-1} = -(\Delta l)^2 \rho_{i,j} \qquad (5.101)$$

or

$$\mathbf{A}_1 \cdot \mathbf{v} + \mathbf{A}_2 \cdot \mathbf{w} = \mathbf{b} \qquad (5.102)$$

The matrix $\mathbf{A}_1$ now acts exclusively on the "rows" of the $u_{i,j}$ lattice, while $\mathbf{A}_2$ effects the "columns" only (see Fig. 5.19). The advantage of equ. 5.102 over 5.89 is that the matrices $\mathbf{A}_1$ and $\mathbf{A}_2$ are tridiagonal, and not pentadiagonal as the matrix $\mathbf{A}$. They may therefore be treated by the fast recursion method of Section 2.1.4.
    The ADI method, then, consists in the iteration of the following double step:

---

**ADI method:**

$$(\mathbf{A}_1 + \omega\mathbf{I}) \cdot \mathbf{v}^{n+1/2} = \mathbf{b} - (\mathbf{A}_2 \cdot \mathbf{w}^n - \omega\mathbf{v}^n) \qquad (5.103)$$
$$\mathbf{w}^{n+1/2} = \mathbf{U} \cdot \mathbf{v}^{n+1/2} \qquad (5.104)$$
$$(\mathbf{A}_2 + \omega\mathbf{I}) \cdot \mathbf{w}^{n+1} = \mathbf{b} - (\mathbf{A}_1 \cdot \mathbf{v}^{n+1/2} - \omega\mathbf{w}^{n+1/2}) \qquad (5.105)$$

---

Here, the optimal value of the relaxation parameter is given by

$$\omega = \sqrt{\lambda_1 \lambda_2}, \tag{5.106}$$

where $\lambda_1$ and $\lambda_2$ are the smallest and largest eigenvalue, respectively, of the matrix $\mathbf{A}$. In the specific case of the potential equation, assuming a lattice with $M = N$, we have $\omega \approx \pi/N$.

EXERCISE: Apply the ADI method to the Laplace problem with $M = N = 5$.

### 5.3.3    Fourier Transform Method (FT)

We consider once more the discretized Poisson equation on a $M \times N$ lattice. This time it is more convenient to enumerate the grid points starting with index 0, i.e. according to $x_k$ ($k = 0, 1, \ldots M - 1$) and $y_l$ ($l = 0, 1, \ldots N - 1$). If the given boundary conditions are compatible with a *periodic* spatial continuation of the basic pattern, meaning that $u_{0,l} = u_{M,l}$ and $u_{k,0} = u_{k,N}$, we may employ the Fourier series representation (see Appendix B)

$$u_{k,l} = \frac{1}{MN} \sum_{m=0}^{M-1} \sum_{n=0}^{N-1} U_{m,n} e^{-2\pi i\, km/M} e^{-2\pi i\, nl/N} \tag{5.107}$$

with

$$U_{m,n} = \sum_{k=0}^{M-1} \sum_{l=0}^{N-1} u_{k,l} e^{2\pi i\, km/M} e^{2\pi i\, nl/N} \tag{5.108}$$

A similar expansion is used for the charge density $\rho_{k,l}$:

$$R_{m,n} = \sum_{k=0}^{M-1} \sum_{l=0}^{N-1} \rho_{k,l} e^{2\pi i\, km/M} e^{2\pi i\, nl/N} \tag{5.109}$$

Inserting these expressions in the equation

$$u_{k+1,l} - 2u_{k,l} + u_{k-1,l} + u_{k,l+1} - 2u_{k,l} + u_{k,l-1} = -(\Delta l)^2 \rho_{k,l} \tag{5.110}$$

we find

$$U_{m,n} = \frac{-R_{m,n}(\Delta l)^2}{2[\cos 2\pi m/M + \cos 2\pi n/N - 2]} \tag{5.111}$$

which may be used in 5.107 to evaluate the solution function $u_{k,l}$. The FT method therefore consists of the steps listed in Figure 5.20. Such a method is competitive only if the numerical Fourier transformation may be performed at a moderate expense in computing time. But this is just what the modern *fast Fourier transform* techniques (FFT; see Appendix B) are offering. To transform $N$ given table values

---

**FT method for periodic boundary conditions:**

- Determine $R_{m,n}$ from

$$R_{m,n} = \sum_{k=0}^{M-1} \sum_{l=0}^{N-1} \rho_{k,l} e^{2\pi i\, km/M} e^{2\pi i\, nl/N} \qquad (5.112)$$

- Compute $U_{m,n}$ according to 5.111

- Insert $U_{m,n}$ in 5.107 to get $u_{k,l}$

---

Figure 5.20: Fourier transform method

they need no more than about $N \ln N$ (instead of $N^2$) operations, and are therefore essential for the considerable success of the FT method.

Boundary conditions other than periodic demand different harmonic expansions. For instance, if the potential values at the boundaries are zero, so that $u_{k,l} = 0$ for $k = 0$, $k = M$, $l = 0$ and $l = N$ (*special*, or *homogeneous* Dirichlet conditions), it is better to use the sine transform of $u$ and $\rho$, defined by

$$u_{k,l} = \frac{2}{M}\frac{2}{N} \sum_{m=1}^{M-1} \sum_{n=1}^{N-1} U^s_{m,n} \sin\frac{\pi km}{M} \sin\frac{\pi ln}{N} \qquad (5.113)$$

$$U^s_{m,n} = \sum_{k=1}^{M-1} \sum_{l=1}^{N-1} u_{k,l} \sin\frac{\pi km}{M} \sin\frac{\pi nl}{N} \qquad (5.114)$$

The function $u$ is then automatically zero at the boundaries. Figure 5.21 gives details of the sine transform procedure.

It turns out that this method may easily be modified so as to cover the case of more general (inhomogeneous) Dirichlet boundary conditions. For instance, let $u$ be given along the lower side of the lattice: $u_{M,l} = u^0_{M,l}$. For the penultimate row $M - 1$ we write

$$u^0_{M,l} - 2u_{M-1,l} + u_{M-2,l} + u_{M-1,l+1} - 2u_{M-1,l} + u_{M-1,l-1} = -(\Delta l)^2 \rho_{M-1,l} \quad (5.117)$$

Subtracting $u^0_{M,l}$ on both sides, we find an equation that is identical to the last of equs. 5.110 for *special* Dirichlet conditions $u_{M,l} = 0$, except that the right-hand hand side now contains an "effective charge density": $\ldots = -(\Delta l)^2 \rho_{M-1,l} - u^0_{M,l}$. Thus we may apply the sine transform method again, using modified charge terms at the boundaries.

*Special* Neumann boundary conditions have the form

$$\left(\frac{\partial u}{\partial x}\right)_{k,l} = \left(\frac{\partial u}{\partial y}\right)_{k,l} = 0 \quad \text{at the lattice boundaries} \qquad (5.118)$$

**FT method for homogeneous Dirichlet boundary conditions ($u = 0$ at the sides):**

- Determine $R^s_{m,n}$ from

$$R^s_{m,n} = \sum_{k=1}^{M-1} \sum_{l=1}^{N-1} \rho_{k,l} \sin \frac{\pi km}{M} \sin \frac{\pi ln}{N} \qquad (5.115)$$

- Compute $U^s_{m,n}$ according to

$$U^s_{m,n} = \frac{-R^s_{m,n}(\Delta l)^2}{2[\cos \pi m/M + \cos \pi n/N - 2]} \qquad (5.116)$$

- Insert $U^s_{m,n}$ in 5.113 to get $u_{k,l}$

Figure 5.21: FT Method using sine transforms

They are most naturally accounted for by a cosine series,

$$u_{k,l} = \frac{1}{2}U^c_{0,0} + \frac{2}{M}\frac{2}{N}\sum_{m=1}^{M-1}\sum_{n=1}^{N-1} U^c_{m,n} \cos \frac{\pi km}{M} \cos \frac{\pi ln}{N} \qquad (5.119)$$

$$U^c_{m,n} = \sum_{k=0}^{M-1}\sum_{l=0}^{N-1} u_{k,l} \cos \frac{\pi km}{M} \cos \frac{\pi nl}{N} \qquad (5.120)$$

For details of the cosine transform method see Figure 5.22.

*General* (inhomogeneous) Neumann boundary conditions of the form

$$\left(\frac{\partial u}{\partial x}\right)_{k,l} = \alpha_{k,l} \quad \left(\frac{\partial u}{\partial y}\right)_{k,l} = \beta_{k,l} \text{ at the lattice boundaries} \qquad (5.124)$$

may again be reduced to *special* Neumann conditions by the introduction of effective charge densities. Writing the last line of the discretized potential equation as

$$u_{M+1,l} - 2u_{M,l} + u_{M-1,l} + u_{M,l+1} - 2u_{M,l} + u_{M,l-1} = -(\Delta l)^2 \rho_{M,l} \qquad (5.125)$$

and requiring that

$$\left(\frac{\partial u}{\partial x}\right)_{M,l} = \alpha_l \qquad (5.126)$$

we approximate the potential on an "outer" line of grid points according to

$$u_{M+1,l} - u_{M-1,l} \approx 2\alpha_l \Delta l \qquad (5.127)$$

---

**FT method for homogeneous Neumann boundary conditions:**

- Determine $R_{m,n}^c$ from

$$R_{m,n}^c = \sum_{k=0}^{M-1}\sum_{l=0}^{N-1} \rho_{k,l} \cos \frac{\pi k m}{M} \cos \frac{\pi l n}{N} \qquad (5.121)$$

- Compute $U_{m,n}^c$ according to

$$U_{m,n}^c = \frac{-R_{m,n}^c (\Delta l)^2}{2[\cos \pi m/M + \cos \pi n/N - 2]} \qquad (5.122)$$

- Insert $U_{m,n}^c$ in

$$u_{k,l} = \frac{1}{2}U_{0,0}^c + \frac{2}{M}\frac{2}{N}\sum_{m=1}^{M-1}\sum_{n=1}^{N-1} U_{m,n}^c \cos \frac{\pi k m}{M} \cos \frac{\pi l n}{N} \qquad (5.123)$$

to find $u_{k,l}$

---

Figure 5.22: FT method using the cosine transform

Subtracting this from 5.125 we find

$$u_{M-1,l} - 2u_{M,l} + u_{M-1,l} + u_{M,l+1} - 2u_{M,l} + u_{M,l-1} = -(\Delta l)^2 \rho_{M,l} - 2\alpha_l \Delta l \quad (5.128)$$

This, however, is identical to the $M$-th line in the case of *special* Neumann conditions $\alpha_l = 0$, except for a modified charge density appearing on the right-hand side. Thus we may again employ the cosine transformation method, using effective charge densities.

## 5.3.4   Cyclic Reduction (CR)

We consider once more the discretized potential equation,

$$u_{k+1,l} - 2u_{k,l} + u_{k-1,l} + u_{k,l+1} - 2u_{k,l} + u_{k,l-1} = -\rho_{k,l}(\Delta l)^2 \qquad (5.129)$$

The grid points are enumerated in the same way as for the FT method: 0 to $N-1$ and $M-1$. For the number of columns in the lattice we choose an integer power of 2: $M = 2^p$. Defining the column vectors

$$\mathbf{u}_k \equiv \{u_{k,l}; \; l = 0, \dots N-1\}^T; \quad k = 0, \dots, M-1 \qquad (5.130)$$

we may write 5.129 as

$$\mathbf{u}_{k-1} \;+\; \mathbf{T} \cdot \mathbf{u}_k + \mathbf{u}_{k+1} = -\rho_k (\Delta l)^2 \qquad (5.131)$$

where

$$\mathbf{T} \;\equiv\; \begin{pmatrix} -2 & 1 & 0 & \cdots \\ 1 & -2 & 1 & \\ & \ddots & \ddots & \ddots \end{pmatrix} - \begin{pmatrix} 2 & 0 & 0 & \cdots \\ 0 & 2 & 0 & \\ & \ddots & \ddots & \ddots \end{pmatrix}$$

$$=\qquad\qquad \mathbf{B} \qquad\qquad - \qquad\qquad 2\,\mathbf{I}$$

Note that $\mathbf{B}$ and $\mathbf{T}$ have the appealing property of being tridiagonal. Next we form linear combinations of every three successive equations 5.131, according to the pattern $[k-1] - \mathbf{T}\cdot[k] + [k+1]$, to find

$$\mathbf{u}_{k-2} + \mathbf{T}^{(1)}\cdot\mathbf{u}_k + \mathbf{u}_{k+2} = -\boldsymbol{\rho}_k^{(1)}(\Delta l)^2 \tag{5.132}$$

with

$$\mathbf{T}^{(1)} \;\equiv\; 2\mathbf{I} - \mathbf{T}^2 \tag{5.133}$$

$$\boldsymbol{\rho}_k^{(1)} \;\equiv\; \boldsymbol{\rho}_{k-1} - \mathbf{T}\cdot\boldsymbol{\rho}_k + \boldsymbol{\rho}_{k+1} \tag{5.134}$$

Evidently, the "reduced" equation 5.132 has the same form as 5.131, except that only every other vector $\mathbf{u}_k$ appears in it. We iterate this process of reduction until we arrive at

$$\mathbf{u}_0 + \mathbf{T}^{(p)}\cdot\mathbf{u}_{M/2} + \mathbf{u}_M = -\boldsymbol{\rho}_{M/2}^{(p)}(\Delta l)^2 \tag{5.135}$$

But the vectors $\mathbf{u}_0$ and $\mathbf{u}_M$ are none other than the given boundary values $u_{0,l}$ and $u_{M,l}$ $(l = 0,1,\ldots N-1)$. Furthermore, the matrix $\mathbf{T}^{(p)}$ is known since it arose from the $p$-fold iteration of the rule 5.133. Of course, $\mathbf{T}^{(p)}$ is not tridiagonal any more; however, it may at least be represented by a $2^p$-fold product of tridiagonal matrices [HOCKNEY 81]:

$$\mathbf{T}^{(p)} = -\prod_{l=1}^{2^p} [\mathbf{T} - \beta_l \mathbf{I}] \tag{5.136}$$

with

$$\beta_l = 2\cos\left[\frac{2(l-1)\pi}{2^{p+1}}\right] \tag{5.137}$$

Thus it is possible to solve 5.135 for the vector $\mathbf{u}_{M/2}$ by inverting $2^p$ tridiagonal systems of equations.

Now we retrace our steps: the vectors $\mathbf{u}_{M/4}$ and $\mathbf{u}_{3M/4}$ follow from

$$\mathbf{u}_0 + \mathbf{T}^{(p-1)}\cdot\mathbf{u}_{M/4} + \mathbf{u}_{M/2} = -\boldsymbol{\rho}_{M/4}^{(p-1)}(\Delta l)^2$$

$$\mathbf{u}_{M/2} + \mathbf{T}^{(p-1)}\cdot\mathbf{u}_{3M/4} + \mathbf{u}_M = -\boldsymbol{\rho}_{3M/4}^{p-1}(\Delta l)^2$$

and so forth.

Hockney has shown that a combination of the CR technique and the Fourier transform method is superior to most other techniques for solving the potential equation [HOCKNEY 70]. In his "FACR" method (for *Fourier analysis and cyclic reduction*) one uses in place of the column vector $\mathbf{u}_k \equiv \{u_{k,l}; \ l = 0, \ldots N - 1\}^T$ its $N$ Fourier components,

$$U_k(n) \equiv \sum_{l=0}^{N-1} u_{k,l} e^{2\pi i n l/N} ; \quad (n = 0, \ldots N - 1) \tag{5.138}$$

Inserting the Fourier series for $u_{k,l}$ in the potential equation one obtains for the $n$-th Fourier component the equation

$$U_{k-1}(n) + U_k(n)\left(2 \cos \frac{2\pi n}{N} - 4\right) + U_{k+1}(n) = -(\Delta l)^2 R_k(n) \tag{5.139}$$

As before, a linear combination of every 3 successive equations may be formed, yielding

$$U_{k-2}(n) \quad + \quad [2 - (2 \cos \frac{2\pi n}{N} - 4)^2] U_k(n) + U_{k+2}(n) =$$

$$= -(\Delta l)^2 [R_{k-2}(n) - (2 \cos \frac{2\pi n}{N} - 4) R_k(n) + R_{k+2}(n)] \tag{5.140}$$

Formal iteration eventually leads to

$$U_0(n) + b^{(p)}(n) U_{M/2}(n) + U_M(n) = -(\Delta l)^2 R_{M/2}^{(p)} \tag{5.141}$$

where $b^{(p)}$ and $\rho_{M/2}^{(p)}$ are given by the iteration. Backwards iteration then yields the desired quantities $U_k(n)$ in succession; inserting them in the Fourier series for $u_{k,l}$ one obtains the solution.

The performance of this method is once again linked to the efficiency of the Fourier transform algorithm. It is therefore absolutely necessary to use the FFT (fast Fourier transform) algorithm explained in Appendix B.

# Part III

# Anchors Aweigh

It is now the time to describe a few applications of the methods developed in Parts I and II. I have tried hard, and failed, to come up with some reasonable categorization of applied computational physics. It seems that computation has transformed *all* parts of physics, and it is probably best to hold on to the usual partitioning of physics into its various branches. Clearly, we cannot cover here all those branches, neatly tracking every possible application of computational techniques. All we can do is discuss a few well-chosen exemplary cases, trying to convey the spirit of the computational approach.

Unabashedly, then, I will start off with my own field of interest, viz. statistical-mechanical simulation. The Monte Carlo technique explained at the beginning of Chapter 6 makes extensive use of the stochastic methods of Chapter 3. In contrast, the "molecular dynamics" method is based on the treatment of classical dynamical equations à la Newton, and the algorithms of Chapter 4 will accordingly play an important role.

Numerical quantum mechanics is a large-scale business, and the large-scale businessmen are mostly chemists, not physicists. However, in addition to the standard program packages of quantum chemistry that, with increasing computer power, are being applied to ever more complex molecules, there are a number of interesting alternative methods tailored to specific problems. Some of these approaches date back to the early days of computer-based stochastics [KALOS 74], while others are relatively new [CAR 85]. Chapter 7 is devoted to an overview of these techniques.

The space-time behavior of flowing continua is described by partial differential equations. Some widely used methods of computational hydrodynamics, obtained by combining the calculus of differences with linear algebra, are explained in Chapter 8.

Another powerful approach to hydrodynamic calculations is based on the concept of a *lattice gas*. This discretized representation of matter may be treated either in the spirit of *cellular automata* propagation or using the newer technique of *Lattice Boltzmann* calculations. Both of these approaches are outlined in Section 8.3.

Finally, the "Direct Simulation Monte Carlo" technique will be described in Section 8.4. It is a widely used method for simulating flow in rarefied gases, be it in aircraft and space engineering or in earthbound gas flow technology.

# Chapter 6

# Simulation and Statistical Mechanics

*Ludwig Boltzmann surely would have approved of simulation*

"Why is the water wet?" says a nursery rhyme in my country. And the grown-up physicist is still striving to explain the macroscopically observable properties of matter in terms of the microscopic kinetics and dynamics of molecules. Since the simultaneous motion of a large number of interacting particles is not tractable by analytical means, statistical mechanics has always been obliged to introduce additional, simplifying assumptions whose effect upon the results is hard to estimate.

What makes the kinetic theory of matter so difficult is not the particularly large number of molecules contained in a chunk or drop of a substance. In fact, the properties of a microdrop of some hundred molecules will differ from those of a macroscopic sample by no more than a few percent. The catch is that we cannot solve, in closed form, the coupled equations of motion of even three particles only, let alone a hundred or more.

However, as soon as computers were available to take over the drudgery of

repetitive calculations, the well-preserved numerical algorithms were brushed up
and applied to various manybody problems.

Incidentally, the term *computer* originally meant just what it says – one who
computes. The earliest computers to actually bear this name were woman em-
ployees of astronomical institutes whose task was the fast and reliable execution
of celestial-mechanical calculations [LANKFORD 90].[1] And the older term "calculator" may be equated to "applied mathematician". It is worth remembering that
none less than Johannes Kepler once held the position of *calculator* ("Rechenmeis-
ter").

Still, it was not before the advent of *electronic* computers that statistical-
mechanical problems could be approached earnestly. In the early years powerful
machines were available only at the American "National Laboratories". Thus the
National Lab at Los Alamos came to be the cradle of statistical-mechanical simulation. Nicholas Metropolis, the Rosenbluths, and Edward Teller employed a
stochastic procedure to sample various configurations of 32 hard disks. Like that
other stochastic method they had developed to treat neutron transport through
matter, they called their technique "Monte Carlo calculation" [METROPOLIS 53].

There existed a prejudice at that time that in a fluid of hard spheres without
attractive pair forces there could not be a solid–liquid phase transition. Thus it
came as quite a surprise when, in the following years, extensive simulations of the
hard disk and hard sphere systems proved the existence of a melting transition
[HOOVER 68].

Only a few years later the *molecular dynamics* method was developed at Law-
rence Livermore Lab. Berni Alder found out that it was feasible to reproduce
by computer simulation the "actual" dynamics going on in a dense fluid of hard
spheres. In a classic paper published in 1957 he formulated the main ingredients
of a workable simulation procedure [ALDER 57]. In the following years he studied
in detail the structural and dynamical properties of the hard sphere fluid. In the
course of these investigations he discovered a very profound and quite unexpected
effect. At low densities – roughly corresponding to the critical density of a real fluid
– there appeared an anomaly of molecule dynamics which Alder and other authors
could later explain as the effect of microscopic vortices. These thermally excited
"Alder vortices", which initially comprise no more than a few dozens of particles,
have the capacity to store part of the momentum a thermally agitated molecule
may possess at some given time, and to gradually pay back the stored momentum
to that molecule. The fluid molecules will thus retain some fraction of their original

---

[1]An amusing example of the early use of "parallel computers" is the development of the first
photographic combination lenses. For this project the Viennese mathematician Petzval, whom
we have encountered before (see page 87), had several artillery men of the Imperial Austro-
Hungarian army (of ranks "Bombardier" and "Oberfeuerwerker") be put under his command. In
the course of the year 1840 these efficient – and, well, sure-fire – calculators traced the paths of
light rays through various lens combinations until an optimum with respect to lens power and
aberrations had been found. In the history of photography the "Petzval lens" has a special place
as the first high-performance objective for portrait work.

velocity for much longer than may be expected according to simple kinetic theories. This may be illustrated in terms of the *velocity autocorrelation function*, which at these densities displays a pronounced "long time tail". A consequence of this is that the mean squared displacement, and consequently the diffusion constant, is far higher than expected.

With Hoover's proof of the existence of a melting transition in hard sphere fluids and Alder's discovery of the *long time tail*, computer simulation rose from its role as a "handmaiden of theory" to an autonomous field of research. In the sixties Aneesur Rahman and Loup Verlet proceeded to perform the first simulations of a Lennard-Jones fluid [RAHMAN 64, VERLET 67]. The interaction potential

$$u_{LJ}(r) = 4\epsilon \left[ (\frac{r}{\sigma})^{-12} - (\frac{r}{\sigma})^{-6} \right] \tag{6.1}$$

(with substance-specific parameters $\epsilon$ and $\sigma$) is richer of detail than the interaction between hard spheres; in fact, it describes rather accurately the forces acting between the atoms in a noble gas. Thus it was possible for the first time to compare the results of simulations to experiments on real substances.

In the years that followed, liquid state physics advanced in great leaps. The microscopic structure and dynamics as well as the thermodynamics and the transport properties of simple fluids were understood ever more clearly. The "Alder vortex" was rediscovered in the Lennard-Jones fluid, again causing an enhanced diffusion coefficient as compared to theoretical predictions [LEVESQUE 69]. The phase transitions solid–liquid [HOOVER 68] and liquid–gas [HANSEN 69] were located, and more recently one could even resolve the long-standing paradox of irreversibility (which apparently should not occur in a classical system obeying reversible equations of motion) [HOLIAN 87, POSCH 90].

In 1971 Aneesur Rahman and Frank H. Stillinger undertook to simulate so complex a liquid as water [RAHMAN 71]. Since then many different model potentials for water have been proposed and used in simulations [NEUMANN 86]. Most of the properties of water and aequeous solutions are by now well understood, while others – mostly those connected to the H-bond structure and to quantum effects – remain fuzzy. With increasing power of the computing machines, but also with increasing refinement of the algorithms, ever more complex molecules became accessible to simulation. In these days program packages are offered that will at the push of a button reproduce the conformational dynamics of proteins made up of several hundred atomic groups [VAN GUNSTEREN 84, BROOKS 83, MACKERRELL 98, SMITH 96, SMITH WWW]. Also, stunning numbers of particles may be followed by simulation. When even the flow patterns in mesoscopic vortices are now computed by molecular simulation [RAPAPORT 88], the borderline towards hydrodynamic phenomena in the strict sense has been crossed.

Various methodological paths have been tried out with the objective of using the available computing power most efficiently. Apart from the molecular dynamics method, the technique of "stochastic dynamics" is often employed. In many instances one is interested only in the motion of a minority of "primary" particles within a large system. An important example is a dilute solution of ions, in which the solvent molecules may be regarded as "extras" whose role is just to provide frictional hindrance as well as thermal agitation to the ions. This type of ionic motion in a viscous, thermally fluctuating medium is reasonably well described by a generalized Langevin equation. One may therefore simulate the ion dynamics by solving this stochastic equation of motion, without having to follow the motion of the far too many solvent particles (see Section 6.6).

An extensive discussion of the the various statistical-mechanical simulation methods and their application would be outside the scope of this book. Suffice it to cite a few out of the many textbooks on this subject: [VESELY 78], in German, is by now somewhat outdated with regard to applications but still valid as an introduction to the basic simulation procedures. [ALLEN 90], in English, is a more recent, excellent methodological overview. Up-to-date reviews of applications of the MC method appear frequently and may be found easily by a web or library search.

In the following sections a coherent sequence of "projects" will serve to provide you with a working knowledge of the two basic simulation methods, Monte Carlo and molecular dynamics. Small, re-usable code units will be developed that may be assembled into complete simulation programs of both kinds.

# 6.1   Model Systems of Statistical Mechanics

## 6.1.1   A Nutshellfull of Fluids and Solids

Simulation requires a model in which the microscopic constituents of a piece of matter are correctly represented. A *fluid*, for one, may be regarded as a collection of atoms or molecules which, if only they are massive enough, will obey the laws of classical mechanics. These particles may then be treated as mass points or rigid bodies interacting with each other by pair forces, and possibly torques, derived from certain model potentials. A list of the most popular interaction potentials is presented in tables 6.1 and 6.2.

A microscopic snapshot of a small subvolume in a simple fluid sample, containing $N$ point particles, is uniquely described by the $N$ positional vectors. If the motion of the particles – in the context of a molecular dynamics simulation – is to be followed, the momentary velocities of all particles must be given as well.

If the position vectors of the $N$ atoms are combined into a vector $\Gamma_c \equiv \{\mathbf{r}_1 \dots \mathbf{r}_N\}$, then the set of all possible such vectors spans the $3N$-dimensional "configuration space" $\Gamma_c$. Given some property $a(\Gamma_c)$ of the $N$-body system, depending on the positions of all particles (i.e. of the *microstate* $\Gamma_c$), the thermodynamic average of the quantity $a$ is given by

$$\langle a \rangle = \int_{\Gamma_c} a(\Gamma_c)\, p(\Gamma_c)\, d\Gamma_c \tag{6.2}$$

where $p(\Gamma_c)$ is the probability density at the configuration space point $\Gamma_c$.

| Hard spheres | $u(r) = \infty$ if $r < r_0$ <br> $= 0$ if $r \geq r_0$ | First approximation in many applications |
|---|---|---|
| Lennard-Jones | $u(r) = 4\,\epsilon\left[\left(\dfrac{r}{\sigma}\right)^{-12} - \left(\dfrac{r}{\sigma}\right)^{-6}\right]$ | Noble gas atoms; near-ly spherical molecules |
| Isotropic Kihara | $u(r) =$ <br> $4\,\epsilon\left[\left(\dfrac{r-a}{\sigma-a}\right)^{-12} - \left(\dfrac{r-a}{\sigma-a}\right)^{-6}\right]$ | Noble gas atoms; near-ly spherical molecules $(a = 0.05 - 0.1\,\sigma)$ |
| Harmonic | $u(r) = A\,(r - r_0)^2$ | Intramolecular bonds, if $kT$ is small compared to the bond energy |
| Morse | $u(r) =$ <br> $A\left[e^{-2b(r-r_0)} - 2e^{-b(r-r_0)}\right]$ | Intramolecular bonds, if $kT$ is comparable to the bond energy |
| Born-Huggins-Mayer | $u(r) =$ <br> $\dfrac{q_1 q_2}{4\pi\epsilon_0 r} + Be^{-\alpha r} - \dfrac{C}{r^6} - \dfrac{D}{r^8}$ | Ionic melts; $q_i$ are the ion charges |

Table 6.1: Isotropic model potentials in statistical-mechanical simulation: $u = u(r)$

| Hard dumbbells, hard spherocylinders, etc. | $u(12) = \infty$   if overlap <br> $\quad\quad\; = 0$   otherwise | First approximation to rigid molecules |
|---|---|---|
| Interaction site models, rigid | $u(12)$ = sum of isotropic pair energies $u(r_{i(1),j(2)})$, where several interaction sites $i$ and $j$ are in fixed positions on molecules 1 and 2, respectively | Rigid molecules |
| Interaction site models with non-rigid bonds | $u(12)$ = sum of isotropic pair energies, both intra- and intermolecular | Non-rigid molecules |
| Kramers-type | $u(12)$ = sum of isotropic pair energies, exclusively between sites on different molecules; certain intramolecular distances (bonds) and/or angles are fixed | Flexible molecules, from ethane to biopolymers |
| Stockmayer | $u(12)$ = Lennard-Jones + point dipoles | First approximation to small polar molecules |
| Anisotropic Kihara | $u(12) =$ <br> $4\epsilon\left[\left(\dfrac{\rho_{12}}{\sigma}\right)^{-12} - \left(\dfrac{\rho_{12}}{\sigma}\right)^{-6}\right]$ <br> where $\rho_{12}$ is the shortest distance between two linear rods | Rigid linear molecules with distributed Lennard-Jones interaction |
| Gay-Berne | $u(12) =$ <br> $4\,\epsilon(12)\left[\left(\dfrac{r_{12} - \sigma(12) + \sigma_0}{\sigma_0}\right)^{-12}\right.$ <br> $\left. - \left(\dfrac{r_{12} - \sigma(12) + \sigma_0}{\sigma_0}\right)^{-6}\right]$ <br> where $\sigma(12)$ and $\epsilon(12)$ depend on $\mathbf{r}_{12}, \mathbf{e}_1, \mathbf{e}_2$ and certain substance-specific shape parameters | Liquid crystal molecules of ellipsoidal shape, with smoothly distributed Lennard-Jones sites |

Table 6.2:   Anisotropic model potentials in statistical-mechanical simulation: $u(12) = u(\mathbf{r}_{12}, \mathbf{e}_1, \mathbf{e}_2)$

It would be all too nice if we could actually compute averages of this form, since the macroscopically measurable properties of a substance are indeed equal to such averages. For instance, it is easy to show that the *internal energy* of a piece of matter is given by

$$U = NkT + \frac{1}{2} \langle \sum_i \sum_{j \neq i} u(r_{ij}) \rangle \qquad (6.3)$$

where $u(r_{ij})$ is the potential energy of a pair of particles with pair distance $r_{ij}$. Similarly, the *pressure* is

$$p = \frac{NkT}{V} - \frac{1}{6V} \langle \sum_i \sum_{j \neq i} r_{ij} \frac{du}{dr} \Big|_{r_{ij}} \rangle \qquad (6.4)$$

The problem with evaluating the expression 6.2 is – apart from the truly high dimensionality of the integral – that the probability density $p(\mathbf{\Gamma}_c)$ is in general unknown. We do know that for instance in the canonical ensemble $p(\mathbf{\Gamma}_c)$ is proportional to the Boltzmann factor $exp[-E(\mathbf{\Gamma}_c)/kT]$, but the normalizing factor $Q$, which according to

$$1 = \int p(\mathbf{\Gamma}_c) \, d\mathbf{\Gamma}_c \equiv \frac{1}{Q} \int e^{-E(\mathbf{\Gamma}_c)/kT} \, d\mathbf{\Gamma}_c \qquad (6.5)$$

defines the absolute value of the probability density, is not known. Incidentally, $Q$ is called the *configurational partition function*.

In a basic model of *ferromagnetic solids* the atoms are taken to reside at fixed positions on the vertices of some appropriate crystal lattice. However, they are carrying dipole vectors (spins) with individually varying directions. In the framework of the early *Ising* model the spins may point either up or down, while the later *Heisenberg* model permits all directions. The microscopic configuration $\mathbf{\Gamma}_c$ of such a model system is defined, not by the (trivial) *positions*, but by the $N$ *spins* on the lattice.

In a two-dimensional square Ising lattice only the four nearest spins are assumed to contribute to the energy of some spin $\sigma_i$ ($= \pm 1$); in three dimensions the six nearest neighbors must be considered. The total energy of the $N$ spins is given by

$$E = -\frac{A}{2} \sum_{i=1}^{N} \sum_{j(i)=1}^{4 \text{ or } 6} \sigma_i \sigma_{j(i)} \qquad (6.6)$$

($A$ being a coupling constant). This expression for the energy may be inserted in the Boltzmann factor to yield the density in canonical phase space.

One relevant "observable" $a\,(\mathbf{\Gamma}_c)$ whose average may be compared to measurements on real ferromagnets is the magnetic polarization

$$M \equiv \sum_{i=1}^{N} \sigma_i \qquad (6.7)$$

as a function of temperature. An external magnetic field $H$ may be applied, with the additional potential energy being given by $E_H = -H \sum_i \sigma_i$.

## 6.1.2    Tricks of the Trade

A few preliminary tasks have to be performed before the actual simulation of a disordered fluid or a spin system may begin. First of all, we have to choose a suitable set of *units* in which to express the mechanical relations; the use of meters, kilograms and such would imply the clumsy manipulation of very small numbers. Next, a reasonable rule must be invented to treat the *boundaries* of our - necessarily quite small - model system. And a suitable *initial configuration* has to be set up from which to start the simulation, and by *adjusting density and temperature* we define a thermodynamic state at which our simulation is to take place.

### Units:

Meters were not meant to measure molecules. It is wise to choose the units of energy, mass, and length such that the values of the mechanical quantities are always in a convenient range, i.e. of order 1. For instance, in simulations with *Lennard-Jones* molecules it is best to count energies in multiples of $E_0 = \epsilon$ and lengths in units of $l_0 = \sigma$. The energy of a given pair of particles is then given by

$$u_{LJ}^* = 4 \left[ r^{*-12} - r^{*-6} \right] \tag{6.8}$$

where $u^* \equiv r/\epsilon$ and $r^* \equiv r/\sigma$. Here we see an additional advantage of using self-consistent or *reduced* units: the Lennard-Jones parameters never occur in the formulae, and we are spared many computationally expensive multiplications along the way. Only after finishing the simulation do we transform the results to the usual metric units, to compare them to experimental data.

Mechanics requires three independent units, of which we have mentioned two. Choosing the atomic mass $m_0 = 1AMU = 1.6606 \cdot 10^{-27} \, kg$ as the third, we arrive at a closed system of mechanical units. The time, which in macroscopic mechanics is one of the basic quantities having a unit of its own, is now measured self-consistently in multiples of $t_0 = \sqrt{m_0 \sigma^2 / \epsilon}$.

If *electrical charges* are in the game, the natural unit to be used is, of course, the electron charge, $q_0 = 1.602 \cdot 10^{-19} \, As$.

The *number density* in a molecular system is simply $\rho = N/V$. This is normally a large number, and again we want to reduce its numerical value by scaling it by some standard density. For Lennard-Jones systems this standard is $\rho_0 = 1/\sigma^3$, and the reduced density is thus defined as $\rho^* \equiv N\sigma^3/V$. In systems of hard spheres of diameter $d_0 = 2 r_0$ the accepted standard density is $\rho_0 = \sqrt{2}/d_0^3$, and the reduced density is thus $\rho^* = Nd_0^3/V\sqrt{2}$.

*Temperature* is best reduced by $T_0 = \epsilon/k$ in the case of Lennard-Jones particles. For hard spheres there is no "natural" unit of energy. It is then convenient to choose $E_0 = kT$ which suggests a self-consistent time unit $t_0 = \sqrt{m_0 d^2/kT}$.

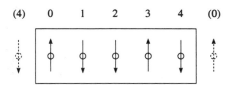

Figure 6.1: Periodic boundary conditions on a spin lattice

## Periodic boundary conditions (PBC) and
## nearest image convention (NIC):

Due to the small size of our model system – typically, $5-100$ molecular diameters
– the majority of fluid particles or lattice spins would be situated near some "wall"
or "boundary", which certainly is not a good representation of the situation inside
a macroscopic sample. Therefore the authors of the very first Monte Carlo studies
already employed "periodic boundary conditions", meaning that they surrounded
the basic cell containing the $N$ particles by periodic images of itself. In the case of
the very short-ranged spin interaction this means that even the last ("rightmost")
spin in a lattice row feels the effect of a right neighbor – whose spin value is simply
taken to be the same as that of the first (leftmost) spin in that row (see Fig. 6.1).
Similar rules apply at the other boundary lines (or faces) of the grid.

In the case of the disordered model fluid the periodic boundary conditions are
defined by the following rule:

Instead of the $x$-coordinate $x_i$ of some particle the quantity

$$(x_i + 2L) \bmod L \qquad\qquad (6.9)$$

(with $L$ the side length of the cell) is stored; the same goes for $y_i$ and
$z_i$. In this way the number of particles within the basic cell is always
conserved. A particle leaving the cell by crossing the right boundary is
automatically replaced by a particle entering from the left, etc. (Adding
$2L$ before performing the modulo operation only serves to catch any
runaway particles with $x_i < -L$.)

To compute the potential pair energy or the force between two particles $i$ and
$j$ one augments the periodic boundary conditions by the so-called *nearest image
convention*. For example, if the coordinate difference $\Delta x_{ij} \equiv x_j - x_i$ is larger than
$L/2$, then the particle $j$ will be disregarded as an interaction partner of $i$, with
its left image, having coordinate $x_j - L$, taking its place. In practice this means
simply that when calculating $u(r_{ij})$ or similar we use the quantity $\Delta x_{ij} - L$ in place
of $\Delta x_{ij}$. An analogous rule holds for $\Delta x_{ij} \leq -L/2$ and for the other coordinates.

Incidentally, it is advantageous to code the conditions $\Delta x > L/2$ etc. without using
the *if* command. Many modern computers offer the possibility of "vectorization", i.e.
the simultaneous execution of a code command acting on an entire array of variables.

The *if* command, however, is a hindrance for vectorization. It is therefore recommended to use the equivalent code line

$$\Delta x = \Delta x - L \cdot \text{nint}\left(\frac{\Delta x}{L}\right) \tag{6.10}$$

which may be vectorized. (Here, nint($a$) denotes the rounded value of $a$, i.e. the integer nearest to $a$).

### Starting configuration:

Setting up an initial configuration for the Ising lattice is simple: draw $N$ spin values at random, with equal probabilities for $+1$ and $-1$. For molecules in disordered media the matter is not as straightforward. If we were to sample the initial positions of the particles at random there would be many particle pairs with unphysically small distances. The strong repulsion – e.g. proportional to $r^{-12}$ for Lennard-Jones molecules – would give rise to very high initial energies and forces, and thus to numerical instabilities. It is therefore customary to initially place the molecules onto the vertices of some crystal lattice – face-centered cubic is very popular for isotropic interactions – and to have them intermingle in a longish "thermalization run" before starting on the simulation proper. Since the population number in a cubic cell with face-centered cubic arrangement is $4m^3$, with $m = 1, 2, \ldots$, the literature abounds with particle numbers such as $N = 32, 108, 256, 500$ etc.

### Adjusting density and temperature:

Since the number of particles is usually fixed, the way to arrive at a desired density is to shrink or expand the volume. This is easy: simply scale all coordinates by a suitable factor.

What about the temperature? In MC simulations temperature is a constant parameter, but in molecular dynamics it is a measurable quantity. Kinetic theory tells us that

$$T = m\langle|\mathbf{v}|^2\rangle/3k \tag{6.11}$$

where $k = 1.3804 \cdot 10^{-23}\, J/deg$ is Boltzmann's constant, and $\langle|\mathbf{v}|^2\rangle$ is the averaged square of the particle velocity. In reduced units, this relation reads $T^* = m^*\langle|\mathbf{v}^*|^2\rangle/3$. To arrive at a desired temperature we first take the average of $|\mathbf{v}^*|^2$ over a number of MD steps and thus determine the actual temperature of the simulated system. Then we scale each velocity component of every particle by $\sqrt{T^*_{desired}/T^*_{actual}}$.

It should be kept in mind that $T^*$ is a fluctuating quantity and can therefore be adjusted only approximately.

EXERCISE: To get a feeling for reduced units, consider a pair of Lennard-Jones particles with $\epsilon = 1.6537 \cdot 10^{-21}\, J$ and $\sigma = 3.405 \cdot 10^{-10}\, m$ – these are the accepted values pertaining to Argon. Let the two molecules be situated at a distance of $3.2 \cdot 10^{-10}\, m$ from each other, and calculate the potential energy of this arrangement. Now do the same calculation using $\epsilon$ and $\sigma$ as units of energy and length, respectively. These parameters then vanish from eq. 6.1, and the calculation is done with quantities of order 1.

Using the above values for the energy and length units together with the atomic mass unit, what is the metric value of the self-consistent unit of time? Let one of the particles have a metric speed $v = 500\,m/s$, typical of the thermal velocities of atoms or small molecules. What is the value of $v$ in self-consistent units?

PROJECT MC/MD: As a first reusable module for a simulation program, write a code to set up a cubic box inhabited by $N = 108$ or $32$ particles in a face-centered cubic arrangement. Use your favourite programming language and make the code flexible enough to allow for easy change of volume (i.e. density). Make sure that the lengths are measured in units of $\sigma_{LJ}$. For later reference, let us call this subroutine STARTCONF.

By scaling all lengths, adjust the volume such that the reduced number density becomes $\rho^* = 0.6$.

PROJECT MD: Augment the subroutine STARTCONF by a procedure that assigns random velocities to the particles, making sure that the sum total of each velocity component is zero.

PROJECT MC/MD: The second subroutine will serve to compute the total potential energy in the system, assuming a Lennard-Jones interaction and applying the nearest image convention:

$$E_{pot} = \frac{1}{2}\sum_{i}\sum_{j\neq i} u_{LJ}(r_{ij}) = \sum_{i=1}^{N-1}\sum_{j=i+1}^{N} u_{LJ}(r_{ij}) \tag{6.12}$$

Write such a subroutine and call it ENERGY. Use it to compute the energy in the system created by STARTCONF.

## 6.2   Monte Carlo Method

In Section 3.3.5 we learned how to compute averages even if the probability density is known no better than up to an undetermined normalization factor. In the context of statistical mechanics this is a well-known problem: the configurational partition function $Q_c$ is in most cases unknown. The trick is to generate a Markov chain of, say, $K$ configurations $\{\mathbf{\Gamma}_c(k),\ k = 1,\dots K\}$ such that the *relative frequency* of a configuration in the chain is proportional to the corresponding Boltzmann factor. We may then estimate the mean value $\langle a \rangle$ from

$$\langle a \rangle = \frac{1}{K}\sum_{k=1}^{K} a\left[\mathbf{\Gamma}_c(k)\right] \tag{6.13}$$

A widely used prescription for generating a suitable Markov chain of microstates is the *biased random walk* through configuration space described in Figure 6.2. The parameter $d$ should be adjusted empirically in such a way that in step 3 approximately one out of two attempted steps $\Gamma'_c$ is accepted.

Incidentally, the random variate sampled in step 1 need not come from an equidistribution; any probability density that is symmetrical about 0, such as a Gauss distribution, will serve the purpose.

Step 3 is the proper core of the MC method. In the case of hard disks or spheres it looks slightly different. $E(k)$ and $E'$ may then assume the values 0 and $\infty$ only, and the Boltzmann factors are either 1 or 0. Figure 6.3 shows the accordingly modified part of the MC procedure.

Still another modification is needed to deal with Ising (or related) systems. The appropriate procedure is described in Figure 6.4.

The basic recipes explained above should be sufficient to guide the reader in writing an Ising MC program and do "experiments" with it.

PROJECT MC (FLUID): Write a subroutine MCSTEP which performs the basic Monte Carlo step as described in Fig. 6.2: selecting at random one of the LJ particles that were placed on a lattice by STARTCONFIG, displace it slightly and apply the PBC; then compute the new potential energy (using NIC!) and check whether the modified configuration is accepted or not, given a specific temperature $T^*$; if accepted, the next configuration is the modified one, otherwise the old configuration is retained for another step.

Write a main routine to combine the subroutines STARTCONF, ENERGY, and MCSTEP into a working MC program.

*Note:* The maximum displacement of a particle in the MC trial move is up to you. It should be chosen such that about 50% of the trial steps are accepted.

PROJECT MC (LATTICE): Let $N = n.n$ spins $\sigma_i = \pm 1$; $i = 1, \ldots N$ be situated on the vertices of a two-dimensional square lattice. The interaction energy is defined by

$$E = \sum_i E_i = -\frac{1}{2} \sum_{i=1}^{N} \sum_{j=1}^{4} \sigma_i \sigma_j \qquad (6.16)$$

where the sum over $j$ involves the 4 nearest neighbors of spin $i$. Periodic boundary conditions are assumed

- Write a Monte Carlo program to perform a *biased random walk* through configuration space.

- Determine the mean total moment $\langle M \rangle \equiv \langle \sum_i \sigma_i \rangle$ and its variance as a function of the quantity $1/kT$. Compare your results to literature data [BINDER 87].

---

**Metropolis Monte Carlo:**

Let $\mathbf{\Gamma}_c(k) \equiv \{\mathbf{r}_1 \ldots \mathbf{r}_N\}$ be given; the potential energy of this configuration is
$E(k) \equiv (1/2) \sum_i \sum_j u(|\mathbf{r}_j - \mathbf{r}_i|)$.

1. Generate a "neighboring" configuration $\mathbf{\Gamma}'_c$ by randomly moving one of the $N$ particles within a cubic region centered around its present position:

$$x'_j \;=\; x_j + d\,(\xi - 0.5)$$

and similarly for $y_j, z_j$. Here, $d$ (= side length of the displacement cube) is a parameter to be optimized (see text), and $\xi$ is a random number from an equidistribution in $(0,1)$. The number $j$ of the particle to be moved may either be drawn among the $N$ candidates, or may run cyclically through the set of particle indices.

2. Determine the modified total energy $E'$; since displacing particle $j$ affects only $N-1$ of the $N(N-1)/2$ pair distances in the system, it is not necessary to recalculate the entire double sum to get $E'$.

3. If $E' \leq E(k)$, we accept $\mathbf{\Gamma}'_c$ as the next element of the Markov chain:

$$E' \leq E(k): \;\Rightarrow\; \mathbf{\Gamma}_c(k+1) = \mathbf{\Gamma}'_c; \;\; \text{go to (1)}$$

If $E' > E(k)$, compare the quotient of the two thermodynamic probabilities,

$$q \;\equiv\; e^{-[E' - E(k)]/kT}$$

to a random number $\xi \in (0,1)$:

$$E' > E(k):$$
$$\begin{aligned} \xi \;&\leq\; q: \;\Rightarrow\; \mathbf{\Gamma}_c(k+1) = \mathbf{\Gamma}'_c; \;\; \text{go to (1)} \\ \xi \;&>\; q: \;\Rightarrow\; \mathbf{\Gamma}_c(k+1) = \mathbf{\Gamma}_c(k); \;\; \text{go to (1)} \end{aligned}$$

(This is the so-called "asymmetric rule"; see also Sec. 3.3.5.)

---

Figure 6.2: Statistical-mechanical Monte Carlo for a model fluid with continuous potential

Let $\Gamma_c(k) \equiv \{\mathbf{r}_1 \ldots \mathbf{r}_N\}$ be given.

- Trial move $\Gamma_c(k) \longrightarrow \Gamma_c'$:

$$x_j' = x_j + d\,(\xi - 0.5) \quad \text{etc., for } y_j,\ z_j \tag{6.14}$$

- If particle $j$ now overlaps with any other particle, let $\Gamma_c(k+1) = \Gamma_c(k)$; otherwise let $\Gamma_c(k+1) = \Gamma_c'$.

Figure 6.3: Monte Carlo for hard spheres

Let $\Gamma_c(k) \equiv \{\sigma_1, \ldots, \sigma_N\}$ be given.

- Pick a spin $\sigma_i$ and tentatively invert it. The resulting energy change is

$$\Delta E = A\sigma_i \sum_{j(i)}^{4} \sigma_j \tag{6.15}$$

- If $\Delta E \leq 0$, accept the inverted spin: $\sigma_i(k+1) = -\sigma_i(k)$; otherwise, draw an equidistributed $\xi \in (0,1)$ and compare it to $w \equiv \exp[-\Delta E/kT]$; if $\xi < w$, accept $-\sigma_i$, else leave $\sigma_i$ unchanged: $\sigma_i(k+1) = \sigma_i(k)$.

Figure 6.4: Monte Carlo simulation on an Ising lattice

# 6.3   Molecular Dynamics Simulation

Two simple examples will serve to demonstrate the principle of the MD method. First we will deal with a system of hard spheres (or disks), then the standard model for simple liquids, the Lennard-Jones fluid, will be treated.

## 6.3.1   Hard Spheres

For an initial configuration of a system of hard spheres we will once again set up a suitable lattice. The $N$ spheres are given random initial velocities, with the additional requirement that the total kinetic energy is to be consistent with some desired temperature according to $E_k = 3NkT/2$. Furthermore, it is advantageous to make the total momentum (which will be conserved in the simulation) equal to zero.

The next step is to find, for each pair of particles $(i, j)$ in the system, the time $t_{ij}$ it would take that pair to meet:

$$t_{ij} = \frac{-b - \sqrt{b^2 - v^2(r^2 - d^2)}}{v^2} \tag{6.17}$$

where $d$ is the sphere diameter, $r$ is the distance between the centers of $i$ and $j$, and

$$
\begin{aligned}
b &= (\mathbf{r}_j - \mathbf{r}_i) \cdot (\mathbf{v}_j - \mathbf{v}_i) \\
v &= |(\mathbf{v}_j - \mathbf{v}_i)|
\end{aligned}
\tag{6.18}
$$

For each particle $i$ the smallest positive collision time $t(i) = min(t_{ij})$ and the corresponding collision partner $j(i)$ is memorized. If particle $i$ has no collision partner at positive times, we set $j(i) = 0$ and $t(i) = [\infty]$, i.e. the largest representable number.

Evidently, the calculation of all possible collision times is quite costly, since there are $N(N-1)/2$ pairs to be scanned. However, this double loop over all indices has to be performed only once, at the start of the simulation.

Next we identify the smallest among the $N$ "next collision times", calling it $t(i_0)$. This gives us the time that will pass until the very next collision occuring in the entire system. Let the partners in this collision be $i_0$ and $j_0$.

Now all particle positions are incremented according to the free flight law

$$\mathbf{r}_i \longrightarrow \mathbf{r}_i + \mathbf{v}_i \cdot t(i_0) \tag{6.19}$$

and all $t(i)$ are decreased by $t(i_0)$.

The elastic collision between the spheres $i = i_0$ and $j = j_0$ leads to new velocities of these two particles:

$$\mathbf{v}'_i = \mathbf{v}_i + \Delta\mathbf{v}, \quad \mathbf{v}'_j = \mathbf{v}_j - \Delta\mathbf{v} \tag{6.20}$$

**Molecular dynamics simulation of hard spheres:**

Immediately after a collision, for each particle $i$ in the system the time $t(i)$ to its next collision and the partner $j(i)$ at that collision is assumed to be known.

1. Determine the smallest positive element $t(i_0)$ among the $t(i)$, identify the corresponding particle $i_0$ and its collision partner $j_0 \equiv j(i_0)$.

2. Let all particles follow their free flight paths for a period $\Delta t \equiv t(i_0)$; subtract $\Delta t$ from each $t(i)$.

3. Perform the elastic collision between $i_0$ and $j_0$; after the collision these spheres have the new velocities

$$\mathbf{v}' - \mathbf{v} \pm \Delta\mathbf{v}, \quad \text{with } \Delta\mathbf{v} = b\frac{\mathbf{r}_{ij}}{d^2} \qquad (6.22)$$

4. Recalculate all times $t(i)$ that involve either $i_0$ or $j_0$, i.e. for $i = i_0$, $i = j(i_0)$, $i = j_0$, and $i = j(j_0)$.

5. Go to (1).

At low densities the large free path may create problems with the periodic boundary conditions, some particle suddenly appearing where it overlaps another. One therefore limits the time allowed for free flight such that for each particle and each coordinate $\alpha$ the free flight displacement fulfills $\Delta x_\alpha \equiv v_\alpha \Delta t \le L/2 - d$.

Figure 6.5: Molecular dynamics of hard spheres

where

$$\Delta\mathbf{v} = b\frac{\mathbf{r}_{ij}}{d^2} \qquad (6.21)$$

All pairwise collision times $t_{ij}$ that involve either $i_0$ or $j_0$ must now be recalculated using the new velocities. For this purpose no more than $2N - 3$ pairs have to be scanned.

The elementary step of a hard sphere MD calculation is now completed. The next step is started by once more searching all $t_i$ for the smallest element. The detailed pattern of a single hard-sphere MD step is described in Fig. 6.5.

EXERCISE: For a two-dimensional system of hard disks, write subroutines to a) set up an initial configuration (simplest, though not best: square lattice;) b) calculate $t(i)$ and $j(i)$; c) perform a pair collision. Combine these subroutines into an MD code. To avoid the difficulty mentioned at the end of Fig. 6.5 one might use reflecting boundary conditions, doing a "billiard dynamics" simulation.

## 6.3.2    Continuous Potentials

The interaction between *hard* particles was treated as an instantaneous collision process, implying forces of infinite strengths acting during infinitely short times. A dynamical equation is of no use in such a model, and it was therefore appropriate to invoke the collision laws for calculating the altered velocities. In contrast, for continously varying pair potentials we have for some particle $i$ at any time $t$

$$\ddot{\mathbf{r}}_i(t) = \frac{1}{m} \sum_{j \neq i} \mathbf{K}_{ij}(t) \tag{6.23}$$

with

$$\mathbf{K}_{ij} \equiv -\nabla_i u(r_{ij}) \tag{6.24}$$

We will consider the standard Lennard-Jones interaction 6.1. For the pair force we find

$$\mathbf{K}_{ij} = -24 \frac{\epsilon}{\sigma^2} \left[ 2 \left( \frac{r_{ij}}{\sigma} \right)^{-14} - \left( \frac{r_{ij}}{\sigma} \right)^{-8} \right] \mathbf{r}_{ij} \tag{6.25}$$

where $\mathbf{r}_{ij} \equiv \mathbf{r}_j - \mathbf{r}_i$.

When evaluating the total force acting on a particle we apply periodic boundary conditions and the *nearest image convention* (see Sec. 6.1). In this way we may determine the quantity on the right-hand side of 6.23. The road is then clear for the stepwise integration of the dynamical equations by one of the methods explained in Chapter 4. One very popular method is the Verlet algorithm

$$\mathbf{r}_i(t_{n+1}) = 2\mathbf{r}_i(t_n) - \mathbf{r}_i(t_{n-1}) + \mathbf{b}_i(t_n)(\Delta t)^2 \tag{6.26}$$

(with $\mathbf{b}_i \equiv \sum_{j \neq i} \mathbf{K}_{ij}/m$). But the predictor-corrector method – usually in the Nordsieck formulation – is also widely used.

PROJECT MD (LENNARD-JONES): Augment the subroutine module ENERGY such that it computes, for each Lennard-Jones particle $i$ in the system, the total force exerted on it by all other particles $j$: $\mathbf{K}_i \equiv \sum_{j \neq i} \mathbf{K}_{ij}$, with $\mathbf{K}_{ij}$ as given by 6.25; remember to apply the *nearest image convention*.

Write a subroutine MOVE to integrate the equations of motion by a suitable algorithm such as 6.26. Having advanced each particle for one time step, do not forget to apply *periodic boundary conditions* to retain them all in the simulation box.

Write a main routine that puts the subroutines STARTCONF, ENERGY and MOVE to work. Test your first MD code by monitoring the mechanically conserved quantities.

Do a number of MD steps – say, 50-100 – and average the quantity $|\mathbf{v}^*|^2$ to estimate the actual temperature. To adjust the temperature to a desired value, scale all velocity components of all particles in a suitable way. Repeat this procedure up to 10 times. After 500-100 steps the fluid will normally be well randomized in space, and the temperature will be steady – though fluctuating slightly.

## 6.3.3    Beyond Basic Molecular Dynamics

There are many generalizations of this basic idea of molecular dynamics simulation, involving orientation dependent potentials and ionic interactions, polymers or other complex molecules. An prerequisite for the simulation of chain molecules is a computationally economic method for treating *geometrical constraints* such as fixed bond length between successive atomic groups in a polymer.

Also, in the many years that have passed since Alder's inspiration we have learned how to simulate nonequilibrium phenomena as well, such as the laminary flow of a liquid. For a detailed discussion of *Non-Equilibrium Molecular Dynamics* (NEMD) we have to refer the reader to the literature, e. g. [EVANS 86] and [HOOVER 99]. However, an important tool in this context is the representation of an *external thermostat* applied to the model system. The most efficient thermostatting recipe, which is of value also in other simulation problems, will be described below [NOSÉ 91].

**Geometrical constraints / SHAKE method:**
In table 6.2 we mentioned model molecules with internal geometrical conditions, such as rigid bond lengths. A straightforward but inefficient method to treat such systems is the introduction of very stiff harmonic bond potentials that permit only small deviations from a given interatomic spacing. The problem with this approach is that large spring constants induce large inaccuracies in the time integration step. To reduce these errors one would have to use a very small time step.

The proven method to accommodate such constraints in a MD program is due to Ryckaert et al. [RYCKAERT 77]. The SHAKE algorithm introduced by these authors is best discussed by considering a system made up of the smallest non-trivial *Kramers chain* molecules consisting of three atoms that are sequentially connected by massless rigid bonds. In each such molecule, the two constraint equations involving the three atomic positions $\mathbf{r}_{1,2,3}$ and the bond lengths $l_{1,2}$ are

$$\sigma_1(\mathbf{r}_{12}) = |\mathbf{r}_{12}|^2 - l_1^2 = 0 \quad \text{and} \quad \sigma_2(\mathbf{r}_{23}) = |\mathbf{r}_{23}|^2 - l_2^2 = 0 \qquad (6.27)$$

where $\mathbf{r}_{12} \equiv \mathbf{r}_2 - \mathbf{r}_1$ etc.

We now postulate Lagrangean constraint forces which, when added to the physical forces acting on each atom, will guarantee the two equations to remain valid as the atoms move around. In the present case it is clear that the constraint force acting on atom 1 must be proportional to $\mathbf{r}_{12}$. Atom 2 takes part in two constraints and therefore is subject to two constraint forces, one directed along $-\mathbf{r}_{12}$ and one along $\mathbf{r}_{23}$. Atom 3 experiences a force pointing along $-\mathbf{r}_{23}$. Thus we may write the three equations of motion

$$\ddot{\mathbf{r}}_1 = \mathbf{b}_1 + \frac{a_1}{m_1}\mathbf{r}_{12} \tag{6.28}$$

$$\ddot{\mathbf{r}}_2 = \mathbf{b}_2 + \frac{1}{m_2}\left[-a_1\mathbf{r}_{12} + a_2\mathbf{r}_{23}\right] \tag{6.29}$$

$$\ddot{\mathbf{r}}_3 = \mathbf{b}_3 - \frac{a_2}{m_3}\mathbf{r}_{23} \tag{6.30}$$

where $\mathbf{b}_{1..3}$ are the physical accelerations due to Lennard-Jones or other pair potentials.

To solve these equations of motion, one proceeds as follows:

1. Given the positions $\mathbf{r}_i^n$ at time $t_n$, integrate the equations of motion for one time step *without considering* the constraint forces; let the resulting positions be denoted as $\mathbf{r}'_i$. Of course, these preliminary position vectors will not fulfill the constraint equations; rather, the values of $\sigma_1(\mathbf{r}'_{12})$ and $\sigma_2(\mathbf{r}'_{23})$ will have some nonzero values which we denote as $\sigma'_1$, $\sigma'_2$.

2. Making the correction *ansatz*

$$\mathbf{r}_1^{n+1} = \mathbf{r}'_1 + \frac{a_1}{m_1}\mathbf{r}_{12}^n \tag{6.31}$$

$$\mathbf{r}_2^{n+1} = \mathbf{r}'_2 + \frac{1}{m_2}\left[-a_1\mathbf{r}_{12}^n + a_2\mathbf{r}_{23}^n\right] \tag{6.32}$$

$$\mathbf{r}_3^{n+1} = \mathbf{r}'_3 - \frac{a_2}{m_3}\mathbf{r}_{23}^n \tag{6.33}$$

and requiring that the corrected position fulfill the constraint equations we have

$$\sigma'_1 - 2\frac{a_1}{\mu_{12}}\left(\mathbf{r}_{12}^n \cdot \mathbf{r}'_{12}\right) + 2\frac{a_2}{m_2}\left(\mathbf{r}_{23}^n \cdot \mathbf{r}'_{12}\right) + [\ldots]^2 = 0 \tag{6.34}$$

$$\sigma'_2 + 2\frac{a_1}{m_2}\left(\mathbf{r}_{12}^n \cdot \mathbf{r}'_{23}\right) - 2\frac{a_2}{\mu_{23}}\left(\mathbf{r}_{23}^n \cdot \mathbf{r}'_{23}\right) + [\ldots]^2 = 0 \tag{6.35}$$

where we have written $1/\mu_{12} \equiv 1/m_1 + 1/m_2$ etc., and where $[\ldots]^2$ are terms that are quadratic in $a_{1,2}$.

We could now solve these two quadratic equations for the unknowns $a_{1,2}$ and insert the solutions in 6.31-6.33 to obtain the corrected positions at time $t_{n+1}$. However, it is more convenient to ignore the small quadratic terms – this is why we have not bothered to write them out explicitly. Rather, the *linear* parts of 6.34-6.35 are solved iteratively, meaning that this system of linear equation is solved to arrive at an improved estimate for $a_{1,2}$ which is again inserted in 6.31-6.33 leading to a new set 6.34-6.35 etc., until the absolute values of $a_{1,2}$ are negligible; generally, this will occur after a very few iterations.

Since we have to iterate anyway, another simplification will do no harm. Instead of solving the linearized equs. 6.34-6.35 exactly at each iteration step, which involves a matrix inversion, we start from one end of the chain and consider only one constraint per atom as we go along. In other words, we first repair the bond $\mathbf{r}_{12}$ by displacing 1 and 2, and then destroy it again by repairing the next bond $\mathbf{r}_{23}$. The point is that by going through the chain several times the errors introduced by neglecting the quadratic terms *and* by considering only one constraint at a time will normally get smaller at each stage.

In our case the procedure is (see 6.34-6.35 without the "cross" terms)

$$a_1 = \frac{\mu_{12}}{2} \frac{\sigma'_1}{\mathbf{r}'_{12} \cdot \mathbf{r}^n_{12}} \qquad a_2 = \frac{\mu_{23}}{2} \frac{\sigma'_2}{\mathbf{r}'_{23} \cdot \mathbf{r}^n_{23}} \tag{6.36}$$

insert this in 6.31-6.33 and iterate until $a_{1,2}$ are negligible.

The generalization of this technique to long chains is trivial. Applications to very long chain molecules, particularly biomolecules, abound in the literature.

**Molecules and robots:**
Another interesting application of "constraint dynamics" may be found in *robotics*. Obviously, robot arms made up of several successive links and joints bear some resemblance to chain molecules. By exploiting this similarity one may attack a standard problem of robotics, known as the *inverse kinematic problem*, in an entirely new way [KASTENMEIER 86].

**Thermostats / Nosé-Hoover method:**
When simulating nonequilibrium processes one is faced with the problem of a gradual temperature rise in the sample. This is not a numerical artifact but a genuine physical effect. The external fields that must be introduced to sustain the nonequilibrium situation necessarily perform work on the system, causing an increase of internal energy.

Introducing a thermostat in a dynamical simulation is a nontrivial task. The temperature of a molecular dynamics sample is not an input parameter to be manipulated at will; rather, it is a quantity to be "measured" in terms of an average of the kinetic energy of the particles,

$$\langle E_{kin} \rangle \equiv \langle \sum_i \frac{mv_i^2}{2} \rangle = d \frac{NkT}{2} \tag{6.37}$$

($d$... dimension). Many authors have come up with suggestions how to maintain a desired temperature in a dynamical simulation – for instance, by repeatedly rescaling all velocities ("brute force thermostat") or by adding a suitable stochastic force acting on the molecules. Such ad hoc tricks have great disadvantages: they are unphysical, and they introduce an artificial trait of irreversibility and/or indeterminacy into the microscopic dynamics. Finally Shuichi Nosé succeeded in finding a thermostating strategy that is compatible with the spirit of microscopic (reversible and deterministic) simulation. Nosé, and later Hoover, could prove that under very mild conditions the following augmented equations of motion will lead to a correct sampling of the canonical phase space at a given temperature $T_0$ :

$$\dot{\mathbf{v}}_i \;=\; \frac{1}{m}\mathbf{K}_i - \xi\,\mathbf{v}_i \tag{6.38}$$

$$\dot{\xi} \;=\; \frac{2}{Q}\,[E_{kin} - 3NkT_0/2] \tag{6.39}$$

In this formulation of the thermostated dynamical equations the coupling parameter $Q$ describes the inertia of the thermostat. The quantity $\xi(t)$ bears some similarity to a viscosity – with the important difference that it is temporally varying and may assume negative values as well.

It should be mentioned that a single NH thermostat produces a thorough perambulation of canonical phase space only if that phase space has more than just a few degrees of freedom. For systems of many particles this is not a problem, but in basic investigations of low-dimensional systems it may be necessary to use two NH thermostats in tandem [MARTYNA 92].

Many profound insights into the foundations of nonequilibrium statistical mechanics have been gained by the application of the deterministic, reversible, yet thermostated equations of motion 6.38-6.39. A more detailed account of the method may be found in [HOOVER 91, HOLIAN 95]. Important applications are given in [POSCH 89, POSCH 92, POSCH 97].

## 6.4   Evaluation of Simulation Experiments

We are now in a position to proceed to calculating averages of the form 6.2. The most elementary thermodynamic observables, pressure and internal energy, may be expressed as averages of the virial and the potential energy, respectively (see equs. 6.3-6.4). The virial is defined by

$$W \equiv \sum_i \mathbf{K}_i \cdot \mathbf{r}_i = -\frac{1}{2}\sum_i\sum_j \mathbf{K}_{ij} \cdot \mathbf{r}_{ij} \tag{6.40}$$

However, the powerful "microscope" of computer simulation gives access to many more details about the structure and the dynamics of statistical-mechanical systems. An important characteristic of microscopic structure is the *pair correlation function* $g(r)$; and the main features of molecular motion are most concisely described in terms of the *velocity autocorrelation function* $C(t)$.

PROJECT MC/MD (FLUID): In your Lennard-Jones MD and MC programs, include a procedure to calculate averages of the total potential energy and the virial. From these compute the internal energy and the pressure. Compare with results from literature, e.g. [MCDONALD 74, VERLET 67]. Allow for deviations in the $5 - 10\%$ range, as we have omitted a correction for the finite sample size ('cutoff correction').

## 6.4.1  Pair Correlation Function

Quite generally, the quantity to be averaged according to equation 6.2 need not be a simple function of dynamical variables; it may well be an "indicator function", or distribution function, of the type

$$a(\mathbf{r}; \mathbf{\Gamma}_c) = \sum_i \delta(\mathbf{r}_i - \mathbf{r}) \qquad (6.41)$$

Averages of this or similar quantities represent relative frequencies — in the present case the relative frequency of some particle residing near $\mathbf{r}$. Such relative frequencies may also be interpreted as *probability densities*. In our example the quantity $\langle a(\mathbf{r}) \rangle = \rho(\mathbf{r})$ would simply denote the mean fluid density at position $\mathbf{r}$:

$$\rho(\mathbf{r}) = \left\langle \sum_i \delta(\mathbf{r}_i - \mathbf{r}) \right\rangle \qquad (6.42)$$

In a fluid we usually have $\rho(\mathbf{r}) = const$; only in the presence of external fields or near surfaces $\rho(\mathbf{r})$ varies in a non-trivial manner. A much more interesting quantity to be evaluated is the "pair correlation function" (PCF)

$$g(r) = \frac{V}{4\pi r^2 N^2} \left\langle \sum_i \sum_{j \neq i} \delta(r - r_{ij}) \right\rangle \qquad (6.43)$$

This is in fact a (ill-normalized) *conditional probability density* – to wit, the probability of finding a particle at $\mathbf{r}$, given that there is a particle at the coordinate origin. $g(r)$, then, provides a measure of spatial ordering in a fluid (or any molecular system).

To determine $g(r)$ in a simulation one first divides the range of $r$-values (at most $[0; L/2]$, where $L$ is the side length of the basic cell) into $50 - 200$ intervals of length $\Delta r$. A given configuration $\{\mathbf{r}_1, \ldots \mathbf{r}_N\}$ is scanned to determine, for each pair $(i, j)$, a channel number

$$k = \text{int}\left(\frac{r_{ij}}{\Delta r}\right) \qquad (6.44)$$

In a histogram table $g(k)$ the corresponding value is then incremented by 1. This procedure is repeated every, say, 50 MD steps (or $50N$ MC steps). At the end of the simulation run the histogram is normalized according to 6.43. The typical shape of the PCF at liquid densities is depicted in Fig. 6.6.

The extraordinary importance of the PCF for the physics of fluids stems from the fact that the average of any quantity that depends on the pair potential $u(r)$ –

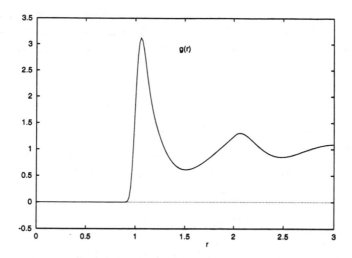

Figure 6.6: Pair correlation function of the Lennard-Jones liquid

and this holds for the majority of physically relevant properties – may be expressed as an integral over $g(r)$. Thus, we have for the pressure (see also 6.4)

$$p = \frac{NkT}{V} - \frac{N^2}{6V^2} \int_V r \frac{du}{dr} g(r) \, d\mathbf{r} \tag{6.45}$$

Moreover, the PCF is the natural meeting place of theory, experiment, and computer simulation. It is possible to compute $g(r)$ for a given pair potential $u(r)$ by analytical means – albeit under rather restrictive simplifying assumptions [KOHLER 72], [HANSEN 86]. And the Fourier transform of $g(r)$, the "scattering law"

$$S(k) = 1 + \frac{N}{V} \int_V [g(r) - 1] \, e^{i\,\mathbf{k}\cdot\mathbf{r}} \, d\mathbf{r} \tag{6.46}$$

is accessible to laboratory experiments. In fact, $S(k)$ is just the relative intensity of neutron or X-ray scattering at a scattering angle $\theta$ which is related to $k$ by

$$k \equiv \frac{4\pi}{\lambda} \sin \frac{\theta}{2} \tag{6.47}$$

PROJECT MD/MC (LENNARD-JONES): Augment your Lennard-Jones MD (or MC) program by a routine that computes the pair correlation function $g(r)$ according to 6.44; remember to apply the *nearest image convention* when computing the pair distances. As the subroutine ENERGY already contains a loop over all particle pairs $(i, j)$, it is best to increment the $g(r)$ histogram within that loop.

Plot the PCF and see whether it resembles the one given in Figure 6.6.

## 6.4.2    Autocorrelation Functions

In *dynamical* simulations not only spatial correlations such as $g(r)$ but also temporal correlations of the type

$$C_a(t) \equiv \langle a(0)a(t) \rangle \tag{6.48}$$

may be computed. An elementary example is the velocity autocorrelation in fluids defined by

$$C(t) \equiv \langle \mathbf{v}_i(0) \cdot \mathbf{v}_i(t) \rangle \tag{6.49}$$

This was the very first autocorrelation function (ACF) to be determined in MD simulations [ALDER 67]. It turned out that at intermediate fluid densities the long time behavior deviates strongly from theoretical expectations. The simplest kinetic theory would predict $C(t) \propto e^{-\lambda t}$; instead, Alder found $C(t) \propto t^{-3/2}$. The diffusion constant $D$ of a liquid is given by

$$D = \frac{1}{3} \int_0^\infty C(t)\, dt \tag{6.50}$$

The value of $D$ is therefore strongly affected by the *long time tail* of $C(t)$; indeed, MD experiments yield values of $D$ that are about 30 percent higher than simple kinetic theory would estimate.

It could later be shown that the surprising persistence of $C(t)$ is due to collective effects. Part of the momentary momentum of a particle is stored in a microscopic vortex that dies off very slowly [DORFMAN 72].

To calculate simple autocorrelation functions in a computer simulation, proceed as follows:

- At regular intervals of $20-100$ time steps, mark starting values $\{a(t_{0,m}), m = 1, \dots M\}$. Since in the further process only the preceding $M \approx 10 - 20$ starting values are required, it is best to store them in registers that are cyclically overwritten.

- At each time $t_n$, compute the $M$ products

$$z_m = a(t_n) \cdot a(t_{0,m}), \quad m = 1, \dots M \tag{6.51}$$

and relate them to the (discrete) time displacements $\Delta t_m \equiv t_n - t_{0,m}$; a particular $\Delta t_m$ defines a channel number

$$k = \Delta t_m / \Delta t \tag{6.52}$$

indicating the particular histogram channel to be incremented by $z_m$. To simplify the final normalization it is recommended to count the number of times each channel $k$ is incremented.

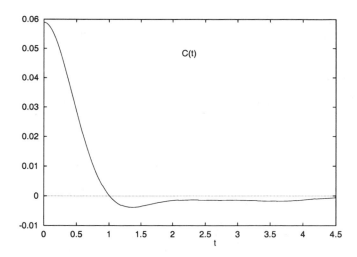

Figure 6.7: Velocity autocorrelation function of the Lennard-Jones fluid

Figure 6.7 shows the general shape of the velocity ACF in a simple fluid.

EXERCISE: Run your MD program for 2000 time steps and store the velocity vector of a certain particle (say, no. 1) at each time step. Write and test a program that evaluates the autocorrelation function of this vector.

PROJECT MD (LENNARD-JONES): Using the experience gathered in the above exercise, write a procedure that computes the velocity ACF, averaged over all particles, during an MD simulation run.
    Plot the ACF and see whether it resembles the one given in Figure 6.7.

## 6.5   Particles and Fields

In describing the basic simulation methods of statistical mechanics we concentrated on interparticle potentials that are negligible beyond a few particle diameters. A measure for the importance of the neglected "tail" of a potential $u(r)$ is the integral

$$\int_{r_{co}}^{\infty} u(r)\, 4\pi r^2\, dr \qquad (6.53)$$

where $r_{co}$ is the cutoff distance. Well-behaved potentials such as Lennard-Jones decay with a high enough negative power of $r$ so as to keep this integral small. There are important cases, however, where we cannot hope for such convenience. The interaction between charged particles decays only as $r^{-1}$, and it will never do to "cut off" the potential at any distance. The same holds for the gravitational potential acting between stars or yet larger assemblies of heavenly matter.

## 6.5.1   Ewald summation

To account for the effect of the long-ranged ion-ion interaction

$$u_{qq} = \frac{q_1 q_2}{r} \tag{6.54}$$

we may take recourse to a method known from solid state theory [EWALD 21]. In the Ewald summation approach the periodic boundary conditions are taken literally: the basic cell containing $N/2$ each of positive and negative charges in some spatial arrangement is interpreted as a single crystallographic element surrounded by an infinite number of identical copies of itself. Such an infinitely extended, globally neutral ion lattice contains an infinite number of charges situated at points $r_{j+}$ and $r_{j-}$, respectively. The total potential at the position of some ion $i$ residing in the basic cell is therefore given by the finite difference of two diverging series:

$$\phi(\mathbf{r}_i) = q \sum_{j+=1}^{\infty} \frac{1}{(\mathbf{r}_i - \mathbf{r}_{j+})} - q \sum_{j-=1}^{\infty} \frac{1}{(\mathbf{r}_i - \mathbf{r}_{j-})} \tag{6.55}$$

The calculation of the potential in $\mathbf{r}$-space would thus lead to an undetermined form $\infty - \infty$. Alternatively, the point charges creating the potential may be described by a sum of delta-like charge densities,

$$\rho(\mathbf{r}) = q \sum_{j+=1}^{\infty} \delta\left(\mathbf{r} - \mathbf{r}_{j+}\right) - q \sum_{j-=1}^{\infty} \delta\left(\mathbf{r} - \mathbf{r}_{j-}\right) \tag{6.56}$$

This periodically varying charge density may be expanded in a Fourier series whose terms determine the Fourier components $\phi(\mathbf{k})$ of the electrostatic potential. In principle these components can be summed to give the total potential at some position. However, the Fourier representation of a delta-function requires infinitely many terms, and the Fourier space calculation would again lead to convergence problems.

A way out of this dilemma is to split up the potential in two well-behaved parts, one being represented in $\mathbf{r}$-space and the other in $\mathbf{k}$-space by rapidly converging series. Without restriction of generality we consider a one-dimensional "ion lattice" with a charge distribution as depicted in Figure 6.8. The delta-like point charges (represented by narrow Gaussians) are augmented by Gaussian charge "clouds" of opposite sign,

$$\rho'(\mathbf{r}) = -q_j \left(\frac{\eta^2}{\pi}\right)^{3/2} e^{-\eta^2 (\mathbf{r} - \mathbf{r}_j)^2} \tag{6.57}$$

to form an auxiliary lattice 1. A further lattice (2) is then introduced to compensate the additional Gaussian charges, such that "lattice 1 + lattice 2 = original lattice".

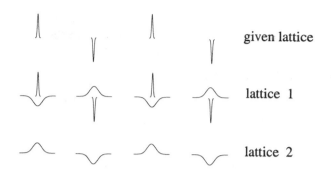

Figure 6.8: Ewald summation

The potential produced by lattice 1 is computed in **r**-space. The farther we walk away from a Gaussian charge cloud, the more it will resemble a delta-like point charge, effectively compensating the original charge it accompanies. Thus the series in **r**-space will converge quite rapidly – the more so if the Gaussians are narrow, i.e. if the parameter $\eta$ in 6.57 is large.

The potential created by lattice 2 is evaluated in **k**-space. If the Gaussians are broad, i.e. if $\eta$ is small, we will need a smaller number of Fourier components. By suitably adjusting $\eta$, optimal convergence of both series may be achieved.

Let us now turn to the more interesting case of three-dimensional model systems. Considering a cubic base cell with side length $L$ containing $N$ charges, the Fourier series now involves the vectors

$$\mathbf{k} \equiv \frac{2\pi}{L}\left(k_x, k_y, k_z\right) \tag{6.58}$$

with integer numbers $k_x$ etc. The most general interparticle vector, involving both base and periodic cell charges, may be written as

$$\mathbf{r}_{i,j,\mathbf{n}} \equiv \mathbf{r}_j + \mathbf{n}L - \mathbf{r}_i \quad (i,j = 1,\dots,N) \tag{6.59}$$

where $\mathbf{n}L$ is a general translation vector in the periodic lattice. Performing the Ewald procedure again we obtain the total potential at position $\mathbf{r}_i$,

$$\phi\left(\mathbf{r}_i\right) = \frac{4\pi}{L^3}\sum_{j=1}^{N}q_j\left[\sum_{\mathbf{k}}e^{-i\mathbf{k}\cdot\mathbf{r}_{ij}}k^{-2}e^{-k^2/4\eta^2} + \sum_{\mathbf{n}}F(\eta|\mathbf{r}_{i,j,\mathbf{n}}|)\right] \tag{6.60}$$

with

$$F(z) \equiv \frac{2}{\sqrt{\pi}}\int_z^{\infty}e^{-t^2}\,dt \tag{6.61}$$

Two tricky details should be mentioned that caused some confusion in the literature before they could be straightened out. First, a Gaussian charge cloud will formally interact with itself, giving rise to a spurious contribution to the potential energy of a point charge $q_i$; this contribution must be subtracted in the final formula. Second, the infinitely repeated lattice should be thought of as the result of a stepwise extension of a finite (roughly spherical) array of image cells. Obviously, the properties of such a finite lattice will depend on the dielectric constant $\epsilon_s$ of the surrounding continuum. It turns out that this influence does not vanish when we take the limit of an infinitely large repeated array. Thus the potential energy of a charge in the base cell contains a contribution from $\epsilon_s$. Usually, one assumes $\epsilon_s = 1$.

Taking into consideration these two corrections, we have for the total potential energy of the system

$$E_{pot} = \frac{1}{2} \sum_{i=1}^{N} q_i \phi\left(\mathbf{r}_i\right) - \frac{\eta}{\sqrt{\pi}} \sum_{i=1}^{N} q_i^2 + \frac{2\pi}{3L^3} \left| \sum_{i=1}^{N} q_i \mathbf{r}_i \right|^2 \qquad (6.62)$$

A similar procedure may be developed for particles that carry point *dipoles* in place of charges. The method is known as "Ewald-Kornfeld summation" This and other methods suited for the dipole-dipole potential, such as the *reaction field* method or Ladd's *multipole expansion method* are explained in [VESELY 78] and [ALLEN 90].

## 6.5.2   Particle-Mesh Methods (PM and P3M):

In large-scale model simulations it is often appropriate not to insist on information about every single constituent particle. Hot plasmas (or galaxies, for that matter) may be described by bunching together some $10^4 - 10^8$ of the ions, electrons, or stars into "superparticles". The position vector of such a superparticle indicates the center of mass of a charge cloud or a cluster of stars. Collisions or interactions between neighboring sub-particles are irrelevant for the behavior of the system as a whole and are therefore neglected. For a detailed discussion of these arguments see [HOCKNEY 81].

The dynamics of a superparticle is governed by the electromagnetic or gravitational field created by all other charges or masses in the system. Due to the long range of these $1/r$-potentials the local field is to a large extent produced by superparticles that are quite far removed from the particle in question. This fact was utilized by Hockney and others to introduce an essential simplification and speed-up of such simulations.

Consider the following model system: a square cell, subdivided into $M \times M$ cells of side length $\Delta x = \Delta y = \Delta l$. The minor cells should still be large enough to contain on the average $10 - 100$ superparticles each. (Taking $M \approx 100$ this means we are dealing with $N \approx 10^5 - 10^6$ superparticles – a formidable number for the molecular dynamicist.) The equation of motion for a superparticle reads

$$\ddot{\mathbf{r}}_k = -\frac{q_k}{m_k} \boldsymbol{\nabla}_k \Phi = \frac{q_k}{m_k} \mathbf{E}(\mathbf{r}_k) \qquad (6.63)$$

where $\Phi(\mathbf{r})$ denotes the solution of the potential equation $\nabla^2 \Phi = -\rho$. The charge (or mass) density $\rho(\mathbf{r}, t)$ is defined by the positions of all superparticles.

Suppose that the configuration of superions is known at some time $t_n$. Our first task is then to compute, using the positions of all particles, the potential function at the centers of the minor cells. The methods explained in Chapter 5.3 are useful here.

The given configuration of superions must first of all be replaced by a discretized, lattice-like charge distribution $\rho_{i,j}$. Various approximations come to mind. The most elementary, called *nearest grid point* (NGP) rule, reads

$$\rho_{i,j} = \frac{1}{(\Delta l)^2} \sum_{k=1}^{N} q_k \, \delta\left(\frac{x_k}{\Delta l} - i\right) \delta\left(\frac{y_k}{\Delta l} - j\right) \qquad (6.64)$$

Here the charge density at the center of each cell $(i, j)$ is determined simply by adding up all charges situated in that cell.

The calculation of the potential may now be performed by a relaxation method or – most efficiently – by the FACR technique as developed by Hockney; see Sec. 5.3.4 and [HOCKNEY 81]. As a result of this step the values of the potential $\Phi_{i,j}$ at the cell centers are available.

Assuming that a given superparticle $k$ is presently located in cell $(i, j)$ we may approximate the field at the position $r_k$ by

$$E_x = -\left[\Phi_{i+1,j} - \Phi_{i-1,j}\right]/2\Delta l \qquad (6.65)$$
$$E_y = -\left[\Phi_{i,j+1} - \Phi_{i,j-1}\right]/2\Delta l \qquad (6.66)$$

Given the local fields, the equation of motion 6.63 may be integrated by a suitable algorithm, such as the Verlet formula

$$\mathbf{r}_k^{n+1} = 2\mathbf{r}_k^n - \mathbf{r}_k^{n-1} + \frac{q_k}{m_k}(\Delta t)^2 \, \mathbf{E}_{i,j}^n \qquad (6.67)$$
$$\mathbf{v}_k^n = \left[\mathbf{r}_k^{n+1} - \mathbf{r}_k^{n-1}\right]/2\Delta t \qquad (6.68)$$

Having thus updated the positions $\mathbf{r}_k^{n+1}$ we may begin the next time step by once again distributing the irregularly located charges to the cell centers and computing the potential $\Phi_{i,j}$. A systematic prescription for the PM procedure is shown in Fig. 6.10.

If the cells are only sparsely inhabited by superparticles, the cell charge $\rho_{i,j}$ changes considerably upon entry or exit of a single particle. The resulting jumps in $\Phi_{i,j}$ and $\mathbf{E}_{i,j}$ tend to destabilize the numerical procedure for integrating the dynamical equations. It is an easy matter to reduce this oversensitivity with respect to charge transfer by applying a more refined method of charge assignment than the NGP rule 6.64. Instead of having all charges contribute with equal weights to the local charge density we distribute appropriate fractions of each charge to the four nearest cell centers. (We are speaking of two dimensions; in three-dimensional systems there would be eight cells in the vicinity.) According to the *cloud in cell* (CIC) rule these fractions, or weights, are assigned in proportion to the overlap areas of a square of side length $\Delta l$, centered around the particle under consideration, and the respective neighbor cells (see Fig. 6.9).

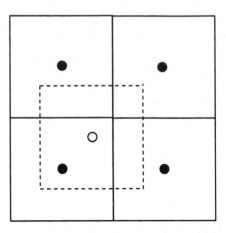

Figure 6.9: Area weighting according to the CIC (cloud-in-cell) rule

**Particle-mesh method:**

At time $t_n$ the spatial distribution $\{\mathbf{r}_k\}$ of the charged (or gravitating) super-particles is given.

1. Assign charge densities $\rho_{i,j}$ to the centers of the cells, either according to the NGP rule

$$\rho_{i,j} = \frac{1}{(\Delta l)^2} \sum_{k=1}^{N} q_k \, \delta\left(\frac{x_k}{\Delta l} - i\right) \delta\left(\frac{y_k}{\Delta l} - j\right) \qquad (6.69)$$

   or by some more refined method such as CIC (see Fig. 6.9).

2. Compute the potential at the cell centers, preferably by the FACR method. For the local field within cell $(i, j)$ use the approximation

$$E_x = -\left[\Phi_{i+1,j} - \Phi_{i-1,j}\right]/2\Delta l \qquad (6.70)$$

   etc.

3. Integrate the dynamical equations up to $t_{n+1}$, for instance by using the Verlet scheme

$$\mathbf{r}_k^{n+1} = 2\mathbf{r}_k^n - \mathbf{r}_k^{n-1} + \frac{q_k}{m_k}(\Delta t)^2 \, \mathbf{E}_{i,j}^n \qquad (6.71)$$

Figure 6.10: Particle-mesh method

In the framework of the PM technique only the fields originating from far-removed superparticles are correctly represented. In many applications the assumption that nearby particles have little influence upon the dynamics is not justified. Be it that we study the interpenetration of galactic spiral arms, investigate the properties of dense plasmas or follow the behavior of ions in melts (or crystals), the short-range interactions must not be neglected.

Similarly, in the simulation of ionic melts by *molecular dynamics* proper – no superparticles, but actual molecules – short-ranged forces are an essential part of the total interaction. One widely used model potential for ions is the one introduced by Born, Huggins and Mayer (see Table 6.1):

$$U(r) = \frac{q_i q_j}{4\pi\epsilon_0 r} + B_{ij} e^{-\alpha_{ij} r} - \frac{C_{ij}}{r^6} - \frac{D_{ij}}{r^8} \tag{6.72}$$

Here we have, in addition to the electrostatic interaction, contributions that are repulsive at short distance (the $B-$ term) and attractive at intermediate distances ($C-, D-$ terms).

Hockney suggested that the optimal strategy in such cases is a mixture of the PM method and the molecular dynamics technique [HOCKNEY 81]. The short-ranged forces are taken into account up to a certain interparticle distance, while the long-ranged contributions are accounted for by the particle-mesh procedure. This combination of particle-particle and particle-mesh methods has come to be called PPPM- or P$^3$M technique.

## 6.6   Stochastic Dynamics

In molecular dynamics experiments we deal with equations of motion of the form

$$\ddot{\mathbf{r}}_i = \frac{1}{m}\sum_j \mathbf{K}_{ij}, \quad i = 1, \ldots N \tag{6.73}$$

By far the most costly step is the evaluation of the $N(N-1)/2$ coupling terms $\mathbf{K}_{ij}$. As a rule some $90-95$ percent of the computing time is spent in the nested loop of the force calculation.

In some applications, however, there are two different classes of degrees of freedom in the system – *primary* ones whose temporal evolution we want to follow, and *secondary* ones that are in fact just dragged along to provide at any given time the complete set of intermolecular forces $\mathbf{K}_{ij}$. The basic example for such a system is a dilute ionic solution of, say, $10-50$ ions in the company of some 5000 water molecules.

In such a situation it may be a good idea to replace the effect of the secondary particles by suitably sampled *stochastic* forces having similar statistical properties as the proper forces $\mathbf{K}_{ij}(t)$.

Forgetting for the moment about the relatively few interactions between ions, we may write down an equation of motion for the single ion in a viscous solvent:

$$\dot{\mathbf{v}}(t) = -\eta \mathbf{v}(t) + \mathbf{a}(t) \tag{6.74}$$

This is Langevin's equation. The statistical properties of the stochastic acceleration $\mathbf{a} \equiv \mathbf{S}/m$ ($\mathbf{S}$... stochastic force) are given by

$$\langle \mathbf{v}(0) \cdot \mathbf{a}(t) \rangle \;\; = \;\; 0 \qquad \qquad \text{for } t \geq 0 \tag{6.75}$$

$$\langle \mathbf{a}(0) \cdot \mathbf{a}(t) \rangle \;\; = \;\; 3 \frac{2\eta kT}{m} \delta(t) \tag{6.76}$$

The first of these relations tells us that $\mathbf{a}(t)$ is not correlated to previous values of the ion velocity; the second equation means that the stochastic and frictional forces are mutually related – which is not surprising since they are both caused by collisions of the ion with solvent molecules. Equation 6.76 gives us only the autocorrelation of the quantity $\mathbf{a}(t)$; the statistical distribution of $|\mathbf{a}|$ is not known a priori. As customary in such cases, we assume that the components of $\mathbf{a}(t)$ are Gauss distributed.

The formal solution to 6.74 reads

$$\mathbf{v}(t) = \mathbf{v}(0)e^{-\eta t} + \int_0^t e^{-\eta(t - t')} \mathbf{a}(t')\, dt' \tag{6.77}$$

By comparing $\mathbf{v}(t)$ (and a corresponding expression for the second integral $\mathbf{r}(t)$) at times $t_n$ and $t_{n+1}$ we find

$$\mathbf{v}_{n+1} \;\; = \;\; \mathbf{v}_n e^{-\eta \Delta t} + \int_0^{\Delta t} e^{-\eta(\Delta t - t')} \mathbf{a}(t_n + t')\, dt' \tag{6.78}$$

$$\mathbf{r}_{n+1} \;\; = \;\; \mathbf{r}_n + \mathbf{v}_n \frac{1 - e^{-\eta \Delta t}}{\eta} + \int_0^{\Delta t} \frac{1 - e^{-\eta(\Delta t - t')}}{\eta}\, \mathbf{a}(t_n + t')\, dt' \tag{6.79}$$

Using the definitions

$$e(t) \equiv e^{-\eta t}, \qquad f(t) \equiv \frac{1 - e^{-\eta t}}{\eta} \tag{6.80}$$

and

$$\mathbf{V}_n \equiv \int_0^{\Delta t} e(\Delta t - t')\, \mathbf{a}(t_n + t') \tag{6.81}$$

$$\mathbf{R}_n \equiv \int_0^{\Delta t} f(\Delta t - t')\, \mathbf{a}(t_n + t') \tag{6.82}$$

the stepwise solution to Langevin's equation may be written as

$$\mathbf{v}_{n+1} = \mathbf{v}_n\, e(\Delta t) + \mathbf{V}_n \qquad\qquad (6.83)$$
$$\mathbf{r}_{n+1} = \mathbf{r}_n + \mathbf{v}_n\, f(\Delta t) + \mathbf{R}_n \qquad\qquad (6.84)$$

The cartesian components of the stochastic vectors $\mathbf{V}_n, \mathbf{R}_n$ are time integrals of the respective components of the $\delta$-correlated stochastic process $\mathbf{a}(t)$ whose statistical properties are given. They are therefore random variates themselves, with statistical properties that are uniquely determined by, and easily derived from, those of the generating process $\mathbf{a}(t)$. In particular, we have $\langle V_n\rangle = \langle R_n\rangle = 0$, $\langle V_n V_{n+1}\rangle = \langle R_n R_{n+1}\rangle = 0$, and

$$\langle V_n^2\rangle = \frac{kT}{m}\left[1 - e^2(\Delta t)\right] \qquad\qquad (6.85)$$
$$\langle R_n^2\rangle = \frac{kT}{m\eta^2}\left[2\eta\Delta t - 3 + 4e(\Delta t) - e^2(\Delta t)\right] \qquad\qquad (6.86)$$
$$\langle V_n R_n\rangle = \frac{kT\eta}{m}\, f^2(\Delta t) \qquad\qquad (6.87)$$

We have learned in Section 3.2.5 how to generate pairs of correlated Gaussian variates. At each time step, then, we may invoke the procedure explained there to produce random numbers $V_n, R_n$ with the desired statistics and insert them, component-wise, in 6.83-6.84.

In the intuitive formulation of equ. 6.74 by P. Langevin, as well as in its much belated stringent derivation, it was always assumed that the stochastic force has a $\delta$-like autocorrelation (see equ. 6.76). This is tantamount to assuming that the solvent particles are much lighter, and therefore faster, than the solute particle. In contrast, if both particle types have comparable masses, the *generalized Langevin equation* applies:

$$\dot{v}(t) = -\int_0^t M(t - t')\, v(t')\, dt' + a(t) \qquad\qquad (6.88)$$

where

$$\langle v(0)a(t)\rangle = 0 \qquad \text{for } t \geq 0 \qquad\qquad (6.89)$$
$$\langle a(0)a(t)\rangle = \frac{kT}{m}\, M(t) \qquad\qquad (6.90)$$

We are now faced with a stochastic *integrodifferential* equation that involves the "history" of the solute particle's motion in the form of the *memory function* $M(t)$ (see [MORI 65]). In practice $M(t)$ is usually fast-decaying, implying that the integrand in 6.88 need be considered for a limited time span only.

There are various methods to render the generalized Langevin equation accessible to numerical work. One group of methods proceeds by approximating the memory function by a certain class of functions. To put it more clearly, one assumes – with good physical justification – that the Laplace transform $\widehat{M}(s)$ may be represented by a truncated chain fraction in the variable $s$. Under this condition the integrodifferential equation may be replaced by a set of coupled differential equations. When written in matrix notation these equations have exactly the same shape as 6.74. They may therefore be treated using the same principles [VESELY 84].

In the other group of techniques one does not attempt to approximate the memory function; instead, one assumes that $M(t)$ may be neglected after $K \approx 20 - 60$ time steps. The random process $a(t)$, whose autocorrelation is given by a limited table of $M(t)$-values, may then be generated as an *autoregressive process* by the method described in Sec. 3.3.3. Replacing the integral in 6.88 by a sum over the most recent $20 - 60$ time steps, one may construct $\mathbf{v}(t)$ and $\mathbf{r}(t)$ in a step-by-step procedure (see [SMITH 90], and also [NILSSON 90]).

# Chapter 7

# Quantum Mechanical Simulation

*Erwin Schrödinger: his equation is at the bottom of it all*

We will not concern ourselves with the time-proven methods that are applied by quantum chemists to compute electronic energies of ever larger molecules; one recommended reference on those crafts is [HEHRE 86]. In the following sections four "physical" techniques will be described that are suited for the investigation of simple quantum systems. They have been applied first to solvated electrons, hydrogen, helium, neon and silicon, and more recently also to metals, carbon and ionic melts.

The technique of *quantum mechanical diffusion Monte Carlo* (QMC, or DMC) dates back to the early days of stochastic simulation. At a meeting held just a few years after publication of the very first statistical-mechanical MC calculations, various ideas on how to treat the Schroedinger equation by stochastic methods were suggested [MEYERS 56]. Many of these ideas were in fact premature, and it took several generations of computing machines before they could be put into action. The "rediscovery" of DMC in the eighties is due to D. Ceperley and – once again – Berni Alder [CEPERLEY 80].

In its basic formulation the DMC method serves to determine the ground state of a bosonic system. The first calculations of this kind were done for $^4$He [KALOS 74, WHITLOCK 79]. Later the method was tuned up in such a way that fermions and excited states may be attacked as well [BARNETT 86, CEPERLEY 88]. Modern applications may be found in [CEPERLEY 96].

With the *path integral Monte Carlo* (PIMC) method we are entering the statistical mechanics of quantum systems. Diffusion Monte Carlo usually refers to the ground state, meaning that the temperature is effectively zero. In PIMC calculations a finite temperature enters by way of a Boltzmann factor – or rather, by its quantum mechanical equivalent, the *density matrix*. Applications of the procedure range from the study of solvated electrons in simple liquids [PARRINELLO 84, COKER 87] to the investigation of the properties of solid para-hydrogen [ZOPPI 91]. Recent PIMC work is surveyed in [CEPERLEY 95] and [CEPERLEY WWW].

*Wave packet dynamics* (WPD) constituted the first attempt of a dynamical semiclassical simulation – an adaptation, as it were, of the molecular dynamics method to quantum mechanics. Building upon ideas proposed by Heller et al. [HELLER 75, HELLER 76], Konrad Singer developed a procedure for simulating the dynamics of "smeared out" neon atoms [SINGER 86]. The further development of the method seems to be possible only by extensive formal and computational effort [HUBER 88, KOLAR 89, HERRERO 95, MARTINEZ 97].

The most exciting new development of the last decades was the designing of a veritable quantum molecular dynamics method by Car and Parrinello [CAR 85]. While in this context the atomic cores (i.e. nucleus plus inner electrons) are still treated as classical particles (Born-Oppenheimer approximation), the outer electrons obey truly quantum mechanical laws. The first substance to be investigated in this manner was amorphous silicon. Subsequently, however, the method has come to be applied to a much wider class of materials: lithium [WENTZCOVICH 91]; microclusters of alkali metals [VITEK 89]; molten carbon [GALLI 90B]; ionic melts [GALLI 90A]. A survey of applications of the technique is given in [VITEK 89].

The dynamical simulation of quantum systems is a rapidly developing branch of computational physics. Many attempts are under way to tame the time-dependent Schroedinger equation, and the dust has not settled sufficiently to show which of these methods will survive. For recent surveys, see [MAKRI 99], [MAZZONE 99], [OHNO 99].

## 7.1   Diffusion Monte Carlo (DMC)

The time-dependent Schroedinger equation for a particle of mass $m$ located in a potential $U(\mathbf{r})$ reads

$$i\hbar\frac{\partial\Psi(\mathbf{r},t)}{\partial t} = H\,\Psi(\mathbf{r},t) \tag{7.1}$$

where the operator $H$ is defined as

$$H \equiv -\frac{\hbar^2}{2m}\nabla^2 + [U(\mathbf{r}) - E_T] \tag{7.2}$$

The *trial energy* $E_T$ is an arbitrary parameter that effects only the – unobservable – phase of the wave function but not its modulus. Introducing a new "imaginary time" variable $s \equiv it/\hbar$ we obtain

$$\frac{\partial \Psi(\mathbf{r}, s)}{\partial s} = D\nabla^2\Psi(\mathbf{r}, s) - [U(\mathbf{r}) - E_T]\,\Psi(\mathbf{r}, s) \tag{7.3}$$

with $D \equiv \hbar^2/2m$.

This equation describes the evolution, in space and "time", of a density $\Psi$ as the consequence of a diffusion process (first term on the right) superposed upon autocatalysis (second term). For easier visualization one may think of a population of bacteria diffusing about in a fluid with locally varying nutrient concentration.

By expanding $\Psi$ in eigenfunctions $\Psi_n$ of the energy operator one may verify the following points:

- If $E_T = E_0$ (ground state energy), then all $\Psi_n$ except $\Psi_0$ will fade out for large "times" $s$:

$$\lim_{s \to \infty} \Psi(\mathbf{r}, s) = \Psi_0(\mathbf{r}) \tag{7.4}$$

- If $E_T > E_0$, the total momentary weight $I(s) = \int \Psi(\mathbf{r}, s)\,d\mathbf{r}$ will grow exponentially in time.

- If $E_T < E_0$, the integral $I(s)$ decreases exponentially in time.

Thus we should try to solve 7.3 for various values of $E_T$, always monitoring the temporal behavior of $I(s)$. If we succeed in finding a value of $E_T$ that gives rise to a solution $\Psi(\mathbf{r}, s)$ whose measure $I(s)$ remains stationary, we may be sure that $E_T = E_0$ and $\Psi = \Psi_0$.

How, then, do we generate a solution to equ. 7.3? Consider the terms on the right-hand side one at a time. The diffusion part of 7.3 reads

$$\frac{\partial n(\mathbf{r}, t)}{\partial t} = D\,\nabla^2 n(\mathbf{r}, t) \tag{7.5}$$

Instead of invoking for this partial differential equation one of the methods of Chapter 5 we may employ a *stochastic* procedure. We have already learned that the diffusion equation is just the statistical summing up of many individual Brownian random walks as described in Sec. 3.3.4. We may therefore put $N$ Brownian walkers on their respective ways, letting them move about according to

$$\mathbf{r}_i(t_{n+1}) = \mathbf{r}_i(t_n) + \boldsymbol{\xi}_i\,, \quad i = 1, \ldots N \tag{7.6}$$

the components $\xi_{x,y,z}$ of the single random step being drawn from a Gauss distribution with $\sigma^2 = 2D\,\Delta t$. If we consider an entire ensemble made up of $M$ such $N$-particle systems, the local distribution density at time $t$,

$$p(\mathbf{r},t) \equiv \langle \delta\,[\mathbf{r}_i(t) - \mathbf{r}] \rangle = \frac{1}{M}\frac{1}{N}\sum_{l=1}^{M}\sum_{i=1}^{N}\delta\,[\mathbf{r}_{i,l}(t) - \mathbf{r}] \qquad (7.7)$$

will provide an excellent estimate for the solution $n(\mathbf{r},t)$ of the diffusion equation 7.5.

At long times $t$ this solution is a very broad, flat and uninteresting distribution, regardless of what initial distribution $n(\mathbf{r},0)$ we started from. However, if there is also a built-in mechanism for a spatially varying autocatalytic process, we will obtain a non-trivial inhomogeneous density even for late times.

The autocatalytic part of the transformed Schroedinger equation 7.3 has the shape

$$\frac{\partial n(\mathbf{r},t)}{\partial t} = f(\mathbf{r})\,n(\mathbf{r},t) \qquad (7.8)$$

Of course, the formal solution to this could be written

$$n(\mathbf{r},t) = n(\mathbf{r},0)\,exp\,[f(\mathbf{r})t] \qquad (7.9)$$

However, we will once more employ a *stochastic* scheme to construct the solution. Again, consider an ensemble of $M$ systems of $N$ particles each. The particles are now fixed at their respective positions; the number $M$ of systems in the ensemble is allowed to vary: those systems which contain many particles located at "favorable" positions where $f(\mathbf{r})$ is high are to be replicated, while systems with unfavorable configurations are weeded out. To put it more clearly, the following procedure is applied when going from $t_n$ to $t_{n+1}$:

- For each of the $M(t_n)$ systems, determine the multiplicity (see equ. 7.9)

$$K_l = exp\left[\sum_{i=1}^{N} f(\mathbf{r}_{i,l})\,\Delta t\right], \quad l = 1,\ldots M(t_n) \qquad (7.10)$$

- Replicate the $l$-th system such that on the average $K_l$ copies are present. To achieve this, produce first int$(K_l)-1$ copies (int$(..) =$ next smaller integer) and then, with probability $w \equiv K_l - \text{int}(K_l)$, one additional copy. (In practice, draw $\xi$ equidistributed $\in [0,1]$ and check whether $\xi \le w$.) If $K_l < 1$, remove, with probability $1 - K_l$, the $l$-th system from the ensemble.

The total number $M(t_n)$ of systems in the ensemble may increase or decrease upon application of this rule. At the end the distribution density 7.7 may again be used to estimate the density at position $\mathbf{r}$.

Let us now apply these ideas to the transformed Schroedinger equation 7.3. Combining the two stochastic techniques for solving the diffusion and autocatalytic equations we obtain the procedure described in Figure 7.1.

**Diffusion Monte Carlo:**

$N$ (non-interacting) particles of mass $m$, distributed at random in a given spatial region, are subject to the influence of a potential $U(\mathbf{r})$. Determine the "diffusion constant" $D = \hbar^2/2m$; choose a trial energy $E_T$, a time step $\Delta s$ and an initial ensemble size $M(s_0)$.

1. For each system $l$ $(= 1, \ldots M(s_0))$ in the ensemble and for each particle $i$ $(= 1, \ldots N)$ perform a random displacement step

$$\mathbf{r}_{i,l}(s_{n+1}) = \mathbf{r}_{i,l}(s_n) + \boldsymbol{\xi}_{i,l}, \quad i = 1, \ldots N \quad (7.11)$$

   where the components of the vector $\boldsymbol{\xi}_{i,l}$ are picked from a Gaussian distribution with $\sigma^2 = 2D\,\Delta s$.

2. For each system $l$ determine the multiplicity $K_l$ according to

$$K_l = exp\left\{\left[\sum_{i=1}^{N} U(\mathbf{r}_{i,l}) - E_T\right]\Delta s\right\} \quad (7.12)$$

3. Produce $int(K_l) - 1$ copies of each system ($int(...)$ denoting the nearest smaller integer;) with probability $w = K_l - int(K_l)$ produce one additional copy, such that on the average there are $K_l$ copies in all. If $K_l < 1$, purge the system with probability $1 - K_l$ from the ensemble.

4. If the number $M$ of systems contained in the ensemble increases systematically (i.e. for several successive steps), choose a smaller $E_T$; if $M$ increases, take a larger $E_T$.

5. Repeat until $M$ remains constant; then the ground state energy is $E_0 = E_T$ and

$$\Psi_0(\mathbf{r}) = \langle\delta(\mathbf{r}_{i,l} - \mathbf{r})\rangle \quad (7.13)$$

Figure 7.1: Quantum mechanical diffusion Monte Carlo

It is evident from the above reasoning that the method will work only for real, non-negative functions $\Psi$. In this basic formulation it is therefore suited only for application to bosonic systems such as $^4$He. Two advanced variants of the technique that may be applied to fermions as well are known as *fixed node* and *released node* approximation, respectively [CEPERLEY 88]. If the node surfaces of $\Psi$ – i.e. the loci of sign changes – are known, then the regions on different sides of these surfaces may be treated separately; within each of these regions $\Psi$ is either positive or negative, and the *modulus* of $\Psi$ is computed by the above method (*fixed node*). Normally the positions of the node surfaces are only approximately known; in such cases they are estimated and empirically varied until a minimum of the energy is found (*released node*).

It must be stressed that the analogy between the wave function $\Psi(\mathbf{r}, t)$ and a probability of residence $n(\mathbf{r}, t)$ which we are exploiting in the DMC method is purely formal. In particular, it has nothing to do with the interpretation of the wave function in terms of a positional probability according to $|\Psi(\mathbf{r})|^2 = prob\{quantum\ object\ to\ be\ found\ at\ \mathbf{r}\}$.

There are situations in which the DMC method in the above formulation is unstable. Whenever we have a potential $U(\mathbf{r})$ that is strongly negative in some region of space, the autocatalytic term in 7.3 will overwhelm everything else, playing tricks to numerical stability. Such problems may be tamed by a modified method called *importance sampling DMC*. Introducing an estimate $\Psi_T(\mathbf{r})$ of the correct solution $\Psi_0(\mathbf{r})$ we define the auxiliary function

$$f(\mathbf{r}, s) \equiv \Psi_T(\mathbf{r})\, \Psi(\mathbf{r}, s) \tag{7.14}$$

By inserting this in 7.3 we find for $f(\mathbf{r}, s)$ the governing equation

$$\frac{\partial f}{\partial s} = D\nabla^2 f - \left[\frac{H\,\Psi_T}{\Psi_T} - E_T\right] f - D\,\nabla \cdot \left[f\,\nabla \ln |\Psi_T|^2\right] \tag{7.15}$$

The autocatalytic term is now small since

$$\frac{H\,\Psi_T}{\Psi_T} \approx E_0 \approx E_T \tag{7.16}$$

The multiplicity $K_l$ will thus remain bounded, making the solution well-behaved.

The last term to the right of equ. 7.15 has the shape of an *advective* contribution. In the suggestive image of a diffusing and procreating bacterial strain it now looks as if there were an additional driving force

$$\mathbf{F}(\mathbf{r}) \equiv \nabla \ln |\Psi_T(\mathbf{r})|^2 \tag{7.17}$$

creating a flow, or drift. The random walk of the individual diffusors has then a preferred direction along $\mathbf{F}(\mathbf{r})$, such that

$$\mathbf{r}_{i,l}(s_{n+1}) = \mathbf{r}_{i,l}(s_n) + \boldsymbol{\xi}_{i,l} + D\Delta s\,\mathbf{F}(\mathbf{r}_{i,l}(s_n)) \tag{7.18}$$

instead of 7.11. And the multiplicity $K_l$ is to be determined from

$$K_l = exp\left\{\left[\frac{H\Psi_T}{\Psi_T} - E_T\right]\Delta s\right\} \tag{7.19}$$

instead of the rule 7.12. All other manipulations described in Fig. 7.1 remain unaltered.

A different formulation of the DMC procedure (actually the older one) is known as "Green's function Monte Carlo" (GFMC); see, among others, [SKINNER 85].

For a recent survey of the DMC method and its applications see [CEPERLEY 96]. Current developments may be followed by visiting respective web sites such as [CEPERLEY WWW] or [CAVENDISH WWW].

# 7.2    Path Integral Monte Carlo (PIMC)

Up to now we have only considered the ground state of an isolated quantum system. Let us now assume that the object of study is part of a larger system having a finite temperature. Then statistical mechanics, in a guise appropriate for quantum systems, enters the stage. Feynman's *path integral* formalism has proved particularly useful in this context.

Since our quantum system is now in contact with a heat bath of temperature $kT > 0$, it must be in a *mixed state* consisting of the various eigenstates of the energy operator:

$$\Psi = \sum_n c_n \Psi_n, \quad \text{where} \quad H\Psi_n = E_n \Psi_n \tag{7.20}$$

The quantum analog of the Boltzmann factor of classical statistical mechanics is the *density matrix* defined by

$$\begin{aligned} \rho(\mathbf{r}, \mathbf{r}'; kT) &\equiv \sum_n \Psi_n^*(\mathbf{r})\, e^{-H/kT}\, \Psi_n(\mathbf{r}') \\ &= \sum_n \Psi_n^*(\mathbf{r})\, e^{-E_n/kT}\, \Psi_n(\mathbf{r}') \end{aligned} \tag{7.21}$$

Writing $\beta$ for $1/kT$, we have for the average of some observable $a(\mathbf{r})$,

$$\langle a \rangle = \int a(\mathbf{r})\, \rho(\mathbf{r}, \mathbf{r}; \beta)\, d\mathbf{r} \bigg/ \int \rho(\mathbf{r}, \mathbf{r}; \beta)\, d\mathbf{r} \equiv Sp[a\rho]\bigg/ Sp[\rho] \tag{7.22}$$

Evidently, the denominator $Sp[\rho]$ here plays the role of a canonical partition function. If we could simply write down $\rho(\mathbf{r}, \mathbf{r}; \beta)$ for a quantum system, the road would be free for a Monte Carlo simulation along the same lines as in the classical case. However, the explicit form of the density matrix is usually quite complex or even unknown. Somehow we will have to get along using only the few simple density matrices we are prepared to handle.

Let us review, therefore, the explicit forms of the density matrix for two very simple models – the free particle and the harmonic oscillator. Just for notational simplicity the one-dimensional case will be considered.

**Density matrix for the free particle:** Let a particle of mass $m$ be confined to a box of length $L$. (We will eventually let $L$ approach $\infty$.) In the absence of an external potential the energy operator reads simply

$$H = -\frac{\hbar^2}{2m} \frac{\partial^2}{\partial x^2} \tag{7.23}$$

and considering the normalization of the eigenfunctions over the interval $[-L/2, L/2]$ we have

$$\Psi_n = \frac{1}{\sqrt{L}} e^{ik_n x}, \quad \text{with} \quad k_n \equiv \frac{2\pi n}{L} \quad \text{and} \quad E_n = \frac{\hbar^2}{2m} k_n^2 \tag{7.24}$$

Inserting this in the definition of the density matrix we obtain

$$\rho_0(x, x'; \beta) = \sum_n \frac{1}{L} e^{-ik_n(x - x')} e^{-\beta \hbar^2 k_n^2 / 2m} \tag{7.25}$$

In the limit $L \to \infty$ the discrete wave number $k_n$ turns into a continuous variable $k$ whose differential is approximated by $dk \approx \Delta k = k_{n+1} - k_n = 2\pi/L$. The sum in $\rho_0$ may then be written as an integral, such that

$$\rho_0(x, x'; \beta) = \frac{1}{2\pi} \int_{-\infty}^{\infty} e^{-ik(x - x')} e^{-\beta \hbar^2 k^2 / 2m} \, dk \tag{7.26}$$

Thus we find for the density matrix of the free particle

$$\rho_0(x, x'; \beta) = \left[ \frac{m}{2\pi \beta \hbar^2} \right]^{1/2} e^{-m(x - x')^2 / 2\beta \hbar^2} \tag{7.27}$$

The probability for the particle to be located at $x$, as given by the diagonal element $\rho_0(x, x; \beta)$, is obviously independent of $x$ – as it must be for a free particle.

**Density matrix for the harmonic oscillator:** A particle of mass $m$ may now be moving in a harmonic potential well,

$$U(x) = \frac{m\omega_0^2}{2} x^2 \tag{7.28}$$

Again determining the eigenfunctions of the energy operator and inserting them in the general expression for the density matrix, we find (see, e.g., [KUBO 71])

$$\rho(x, x'; \beta) \;=\; \left[\frac{m\omega_0}{\pi\hbar}\, \tanh\frac{\beta\hbar\omega_0}{2}\right]^{1/2}$$

$$\cdot exp\left\{-\frac{m\omega_0}{4\hbar}\left[(x + x')^2\, \tanh\frac{1}{2}\beta\hbar\omega_0 + (x - x')^2\, \coth\frac{\beta\hbar\omega_0}{2}\right]\right\} \qquad (7.29)$$

For the evaluation of statistical-mechanical averages we require only the diagonal elements

$$\rho(x, x; \beta) = \left[\frac{m\omega_0}{\pi\hbar}\, \tanh\frac{\beta\hbar\omega_0}{2}\right]^{1/2} exp\left\{-\frac{m\omega_0}{\hbar}x^2\, \tanh\frac{\beta\hbar\omega_0}{2}\right\} \qquad (7.30)$$

The trick in the PIMC method is to express the density matrix of *any* given system in terms of the free particle density 7.27. The following transformation provides an excuse for doing this:

$$\rho(x, x'; \beta) \;=\; \sum_n \Psi_n^*(x)\, e^{-\beta H}\Psi_n(x') =$$

$$=\; \sum_n \Psi_n^*(x)\, e^{-\beta H/2}\, e^{-\beta H/2}\Psi_n(x') =$$

$$=\; \sum_n \Psi_n^*(x)\, e^{-\beta H/2} \int dx'' \delta(x' - x'')\, e^{-\beta H/2}\Psi_n(x'') =$$

$$=\; \sum_n \Psi_n^*(x)\, e^{-\beta H/2} \int dx'' \sum_m \Psi_m(x')\, \Psi_m^*(x'')\, e^{-\beta H/2}\Psi_n(x'') =$$

$$=\; \int dx'' \left[\sum_n \Psi_n^*(x)\, e^{-\beta H/2}\, \Psi_n(x'')\right]\left[\sum_m \Psi_m^*(x'')\, e^{-\beta H/2}\Psi_m(x')\right]$$

Therefore,

$$\rho(x, x'; \beta) = \int dx''\rho(x, x''; \frac{\beta}{2})\, \rho(x'', x'; \frac{\beta}{2}) \qquad (7.31)$$

The expression on the right-hand side is known as a *path integral*. The beauty of it is that the integrand consists of two density matrices pertaining to $\beta/2$, i.e. double the original temperature. But the higher the temperature, the smaller will the effect of the potential $U(x)$ be – and the more closely will the respective factor in the integrand resemble the density matrix of a free particle. Might there be a way to iterate this formal procedure, such that the remaining high-temperature density matrices may essentially be equated to the free particle density $\rho_0$?

There is such a way. Writing $x_0, x_1, x_2 \ldots$ in place of $x, x', x'', \ldots$, we have the strict relation

$$\rho(x_0, x_P; \beta) = \int \ldots \int dx_1 \, dx_2 \ldots dx_{P-1} \, \rho(x_0, x_1; \frac{\beta}{P}) \ldots \rho(x_{P-1}, x_P; \frac{\beta}{P}) \qquad (7.32)$$

The number $P$ of intermediate steps is called the *Trotter number*. If we only choose $P$ large enough – in practice, between 5 and 100 – the following ansatz provides a good approximation to the real thing:

$$\rho(x_p, x_{p+1}; \frac{\beta}{P}) \equiv \sum_n \Psi_n^*(x_p) \, e^{-(\beta/P)\,[H_{free} + U(x)]} \, \Psi_n(x_{p+1})$$

$$\approx \sum_n \Psi_n^*(x_p) \, e^{-(\beta/P)H_{free}} \Psi_n(x_{p+1}) e^{-(\beta/2P)\,[U(x_p) + U(x_{p+1})]}$$

$$= \rho_0(x_p, x_{p+1}; \frac{\beta}{P}) \, e^{-(\beta/2P)\,[U(x_p) + U(x_{p+1})]} \qquad (7.33)$$

For the diagonal element $\rho(x_0, x_0; \beta)$ required to perform averages we find

$$\rho(x_0, x_0; \beta) = A^{P/2} \int \ldots \int dx_1 \ldots dx_{P-1} \, e^{-\beta(U_{int} + U_{ext})} \qquad (7.34)$$

with

$$A \equiv \frac{mP}{2\pi\beta\hbar^2}, \qquad U_{ext} \equiv \sum_{p=0}^{P-1} U(x_p)/P \qquad (7.35)$$

and

$$U_{int} \equiv \frac{A\pi}{\beta} \sum_{p=0}^{P-1} (x_p - x_{p+1})^2 \qquad (7.36)$$

Proceeding now to the more relevant case of three dimensions, we have

$$\rho(\mathbf{r}_0, \mathbf{r}_0; \beta) = A^{3P/2} \int \ldots \int d\mathbf{r}_1 \ldots d\mathbf{r}_{P-1} \, e^{-\beta(U_{int} + U_{ext})} \qquad (7.37)$$

with the same $A$ as above, and

$$U_{ext} \equiv \sum_{p=0}^{P-1} U(\mathbf{r}_p)/P, \qquad U_{int} \equiv \frac{A\pi}{\beta} \sum_{p=0}^{P-1} |\mathbf{r}_p - \mathbf{r}_{p+1}|^2 \qquad (7.38)$$

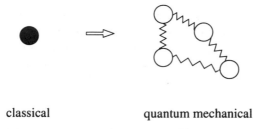

classical                              quantum mechanical

Figure 7.2: Classical isomorphism for one particle

Chandler and Wolynes have pointed out that the expression 7.37 has the shape of a classical Boltzmann factor pertaining to a particular kind of *ring polymer* [CHANDLER 81]. The $P$ elements of this polymer are under the influence of an external potential

$$u_{ext}(x_p) = U(x_p)/P \qquad (7.39)$$

Successive links of the ring chain are coupled together by a harmonic bond potential

$$u_{int}(x_p, x_{p+1}) = \frac{A\pi}{\beta} (x_p - x_{p+1})^2 \qquad (7.40)$$

where we have put $x_P = x_0$. This so-called *classical isomorphism* is illustrated in Figure 7.2.

It follows that we may play the old classical Monte Carlo game to obtain quantum statistical averages. All we have to do is replace a single particle by a flexible ring polymer made up of $5 - 100$ links that are coupled according to the pattern just described. The PIMC procedure for one particle in an external potential is described in Fig. 7.3.

The strength of the springs acting between successive elements of the polymer, as given by $k = Pm/\beta^2\hbar^2$, increases with larger Trotter numbers, while the influence of the external potential will decrease according to $u_{ext}(\mathbf{r}_p) = U(\mathbf{r}_p)/P$. The forceful springs permit only very small displacements per Monte Carlo step, although the mild variation of $u_{ext}(\mathbf{r}_p)$ would allow much larger strides.

This dilemma may be solved by first moving the entire ring polymer, without changing its shape, by a large random step, and subsequently displacing the individual elements relative to each other by a small amount. Another way out is to construct the entire ring polymer anew at each time step, sampling the single element positions from the narrow multivariate Gauss distribution

$$p(\mathbf{r}_0, \dots \mathbf{r}_{P-1}) \propto exp\left[ -\frac{mP}{2\beta\hbar^2} \sum_{p=0}^{P-1} (\mathbf{r}_p - \mathbf{r}_{p+1})^2 \right] \qquad (7.44)$$

and to displace the center of mass of the chain by a wide random step.

**Path integral Monte Carlo for one particle:**

A particle that is under the influence of an external potential $U(\mathbf{r})$ is represented by a ring polymer consisting of $P$ links. Let $\mathbf{r} \equiv \{\mathbf{r}_0, \ldots \mathbf{r}_{P-1}\}$ and the according potential energy

$$U_{pot}(\mathbf{r}) \equiv U_{int}(\mathbf{r}) + U_{ext}(\mathbf{r}) \tag{7.41}$$

be given, with

$$U_{int} \equiv \frac{mP}{2\beta^2\hbar^2} \sum_{p=0}^{P-1} |\mathbf{r}_p - \mathbf{r}_{p+1}|^2 \tag{7.42}$$

$$U_{ext} \equiv \frac{1}{P} \sum_{p=0}^{P-1} U(\mathbf{r}_p) \tag{7.43}$$

1. Displace $\mathbf{r}$ as a whole by $\Delta\mathbf{r}$ (large); also, move each link $\mathbf{r}_p$ by a small amount $\Delta\mathbf{r}_p$; the new configuration is called $\mathbf{r}'$.

2. Compute $U_{pot}(\mathbf{r}')$ and $\Delta U \equiv U_{pot}(\mathbf{r}') - U_{pot}(\mathbf{r})$.

3. Metropolis step: Draw $\xi$ from an equidistribution in $[0, 1]$;
   if $\Delta U \leq 0$, put $\mathbf{r} = \mathbf{r}'$;
   if $\Delta U > 0$ and $\xi \leq e^{-\beta \Delta U}$, put $\mathbf{r} = \mathbf{r}'$ as well;
   if $\Delta U > 0$ and $\xi > e^{-\beta \Delta U}$, let $\mathbf{r}$ remain unchanged.

4. Return to (1).

Figure 7.3: PIMC for one particle

The preceding considerations may without any difficulty be generalized to the case of $N$ particles interacting by a pair potential $u(|\mathbf{r}_j - \mathbf{r}_i|)$. *Each* of these particles has to be represented by a $P$-element ring chain. Denoting the position of element $p$ in chain (=particle) $i$ by $\mathbf{r}_{i,p}$, we have for the diagonal element of the total density matrix

$$\rho(\mathbf{r}_0, \mathbf{r}_0; \beta) =$$
$$A^{3NP/2} \int \ldots \int d\mathbf{r}_{1,1} \ldots d\mathbf{r}_{N,P-1} \, exp\left[ -A\pi \sum_{i=1}^{N} \sum_{p=0}^{P-1} (\mathbf{r}_{i,p} - \mathbf{r}_{i,p+1})^2 \right]$$
$$\cdot exp\left[ -\frac{\beta}{P} \sum_{i=1}^{N} \sum_{j>i} \sum_{p=0}^{P-1} u(|\mathbf{r}_{j,p} - \mathbf{r}_{i,p}|) \right] \quad (7.45)$$

with $\mathbf{r}_0 \equiv \{\mathbf{r}_{1,0} \ldots \mathbf{r}_{N,0}\}$. Obviously, the pair potential acts only between *respective* links ($p$) of *different* chains.

EXERCISE: Write a PIMC program treating the case of one particle of mass $m$ in a two-dimensional oscillator potential $U(\mathbf{r}) = kr^2/2$. Let the Trotter number vary between 2 and 10. Determine the positional probability $p(\mathbf{r})$ of the particle from the relative frequency of residence at $r$, averaged over all chain links. Noting that

$$p(\mathbf{r}) \equiv \rho(\mathbf{r}, \mathbf{r}; \beta) \quad (7.46)$$

we would expect for the two-dimensional harmonic oscillator (with $\omega_0^2 = k/m$)

$$p(r) = 2\pi r \left[ \frac{A}{\pi} \right] e^{-Ar^2}, \quad \text{where } A \equiv \frac{m\omega_0^2}{\hbar} tanh\frac{\beta\hbar\omega_0}{2} \quad (7.47)$$

(For convenience, put $\hbar = 1$.) Draw several configurations of the ring polymer that occur in the course of the simulation.

Three examples may serve to illustrate the practical application of the PIMC method.

Parrinello and Rahman studied the behavior of a solvated electron in molten KCl [PARRINELLO 84]. The physical question here is whether such electrons are localized or smeared out in the quantum manner. (Apart from the theoretical interest of such simple quantum systems, solvated electrons may serve as spectroscopic probes for the microscopic dynamics in polar liquids. And they are an attractive playground for trying out the PIMC method.) The simulation yielded the definitive answer that electrons in molten KCl are clearly localized.

Coker et al. investigated the solvation of electrons in *simple* fluids, in contrast to the molten halogenide studied by Parrinello and Rahman. It turns out that an electron in liquid helium will be strongly localized, whereas in liquid xenon it has quite an extended positional probability (see Fig. 7.4). The probable reason for this behavior is that the atomic shell of He is rather rigid and difficult to polarize,

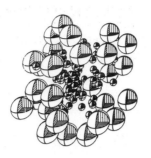

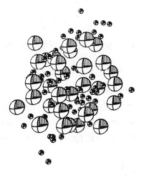

Figure 7.4: From Coker et al.: solvated electron a) in liquid helium, b) in liquid xenon

resulting in a strong repulsion experienced by an extra electron. The solvated particle is therefore surrounded by rigid walls that confine it much like a cage. In contrast, the shells of the larger noble gas atoms are easily polarizable, producing a long-ranged dipole potential that adds up to a flat local potential; the solvated electron is therefore "quasi-free" [COKER 87].

Zoppi and Neumann studied the properties of solid parahydrogen [ZOPPI 91]. The kinetic energy contained in the lattice may be measured by neutron scattering, but is also accessible to PIMC simulation. The authors found good agreement between experiment and simulation. (Due to its small mass, hydrogen is an eminently quantum mechanical system; any attempt to calculate the energy along classical or semi-classical lines is therefore doomed to failure.)

A fairly recent survey of PIMC applications is [CEPERLEY 95]. An update to this paper and other relevant information may be found at [CEPERLEY WWW].

## 7.3    Wave Packet Dynamics (WPD)

Particles of moderately small mass, such as neon atoms, may not be treated as classical point masses, yet do not require a full-fledged quantum mechanical treatment. The quantum broadening is small enough to permit simple approximations. A useful approach is to represent the wave packet describing the (fuzzy) position of the atomic center of some particle $k$ by a Gaussian:

$$\phi_k(\mathbf{r}, t) = e^{\frac{i}{\hbar} Q_k(t)} \tag{7.48}$$

where the quadratic form $Q_k$ is defined by

$$
\begin{aligned}
Q_k(t) &\equiv [\mathbf{r} - \mathbf{R}_k(t)]^T \cdot \mathbf{A}_k(t) \cdot [\mathbf{r} - \mathbf{R}_k(t)] + \mathbf{P}_k(t) \cdot [\mathbf{r} - \mathbf{R}_k(t)] + D_k(t) \\
&\equiv \boldsymbol{\xi}_k^T(t) \cdot \mathbf{A}_k(t) \cdot \boldsymbol{\xi}_k(t) + \mathbf{P}_k(t) \cdot \boldsymbol{\xi}_k(t) + D_k(t) \tag{7.49}
\end{aligned}
$$

The center of the packet, then, is located at $\mathbf{R}_k(t)$. The matrix $\mathbf{A}_k(t)$ describes the momentary shape, size and orientation of the wave packet. In the most simple case $\mathbf{A}_k$ is scalar, making the wave packet spherically symmetric. In general $\mathbf{A}_k$ describes an ellipsoidal "cloud" with a typical size of about $\sigma_{LJ}/10$. The vector $\mathbf{P}_k(t)$ defines the momentum of the wave packet, and $D_k(t)$ is a phase factor that takes care of normalization.

For easy visualization of the formalism let us consider the one-dimensional case. An individual wave packet is described by

$$\phi(x, t) = e^{\frac{i}{\hbar} Q(t)} \tag{7.50}$$

where

$$
\begin{aligned}
Q(t) &= A(t)[x - X(t)]^2 + P(t)[x - X(t)] + D(t) \\
&\equiv A(t)\xi^2(t) + P(t)\xi(t) + D(t) \tag{7.51}
\end{aligned}
$$

($A$ and $D$ are in general complex; $P$ is real.) The expectation value of the position operator $x$ is then given by

$$\langle \phi | x | \phi \rangle \equiv \int dx \, x \, \phi^*(x, t) \phi(x, t) = X(t) \tag{7.52}$$

and the expected momentum is

$$\langle \phi | - i\hbar \frac{\partial}{\partial x} | \phi \rangle = \ldots = P(t) \tag{7.53}$$

Thus the given wave packet indeed represents a semiclassical particle located at $X(t)$ and having momentum $P(t)$.

The assumption of a Gaussian shape for the wave packet has no physical foundation. It is made for mathematical convenience, the argument being that *any* approximation that goes beyond the classical assumption of a mass point (i.e. a $\delta$-like wave packet) will improve matters. The specific advantage of the Gaussian shape as compared to others is that such a wave packet, when subjected to the influence of a *harmonic* potential, will retain its Gaussian shape – albeit with parameters **A**, **P** and $D$ that may change with time. But *any* continuous potential may be approximated locally by a quadratic function, i.e. a harmonic potential.

The wave function 7.48 describes a single particle. In a system of $N$ atoms the complete wave function is often approximated by the product

$$\Psi(\mathbf{r}, t) = \prod_{k=1}^{N} \phi_k(\mathbf{r}, t) \tag{7.54}$$

The effects of exchangeability are thus assumed to be negligible – a safe bet when dealing with medium-mass atoms such as neon.

We are now ready to solve the time-dependent Schroedinger equation

$$i\hbar \frac{\partial \Psi(\mathbf{r}, t)}{\partial t} - H\Psi(\mathbf{r}, t) = 0 \tag{7.55}$$

Following a suggestion of Heller, we apply the minimum principle of Dirac, Frenkel, and McLachlan. The DFM principle tells us that the temporal evolution of the parameters $\mathbf{A}_k$, $\mathbf{P}_k$, and $D_k$ must occur in such a way that the expression

$$I\left(\Psi, \frac{\partial \Psi}{\partial t}\right) \equiv \int \cdots \int \left| i\hbar \frac{\partial \Psi}{\partial t} - H\Psi \right|^2 d\mathbf{r} \tag{7.56}$$

will at all times assume its minimum value.

By applying the tools of variational calculus to this problem, and introducing the simplifying assumption that $\mathbf{A}_k = A_k\mathbf{I}$ (spherical Gaussian) one obtains the following equations of motion for the quantities $A_k$, $\mathbf{P}_k$, and $D_k$ (omitting the particle index $k$):

$$\left(\dot{A} + \frac{2}{m}A^2\right)\langle\xi^2\rangle + \langle\overline{U}\rangle + \left[-\frac{3\hbar i}{m}A - \frac{P^2}{2m} + \dot{D}\right] = 0 \tag{7.57}$$

$$\dot{P}_\alpha\langle\xi_\alpha^2\rangle + \langle\overline{U}\xi_\alpha\rangle = 0 \tag{7.58}$$

$$\left(\dot{A} + \frac{2}{m}A^2\right)\langle(\xi^2)^2\rangle + \langle\overline{U}\xi^2\rangle + \left[-\frac{3\hbar i}{m}A - \frac{P^2}{2m} + \dot{D}\right]\langle\xi^2\rangle = 0 \tag{7.59}$$

Here, $\langle\ldots\rangle$ denotes an expectation value, and

$$\overline{U}_k \equiv \sum_{l \neq k} \int U(r_{kl})\phi_l^*\phi_l \, d\mathbf{r}_l \tag{7.60}$$

is the potential created at $\mathbf{r}_k$ by the "smeared out" particles $l$. Singer et al. recommend to approximate the given pair potential $U(r)$ by a sum of Gaussian functions; in this way the right-hand side of 7.60 can be split up into a sum of simple definite integrals.

The above equations (7.57-7.59) may be cast in a more compact form by introducing auxiliary variables $c$, $d$, and $Z$ according to

$$c \equiv \langle (\xi^2)^2 \rangle - \langle \xi^2 \rangle^2 \tag{7.61}$$

$$d \equiv \langle \overline{U} \xi^2 \rangle - \langle \overline{U} \rangle \langle \xi^2 \rangle \tag{7.62}$$

and

$$A \equiv \frac{m}{2} \frac{\dot{Z}}{Z} \tag{7.63}$$

With $\dot{\mathbf{R}} \equiv \mathbf{P}/m$, the equations of motion for the position $\mathbf{R}$ and the shape parameter $Z$ read

$$\ddot{R}_\alpha = -\frac{\langle \overline{U} \xi_\alpha \rangle}{m \langle \xi_\alpha^2 \rangle} \qquad \ddot{Z} = -\frac{2}{m} \frac{d}{c} Z \tag{7.64}$$

They can be solved using any appropriate integration method, such as the Størmer-Verlet algorithm.

Singer and Smith applied this procedure to liquid and gaseous neon [SINGER 86]. The basic thermodynamic properties could be reproduced in good agreement with experimental values. The pair correlation function exhibits a smearing out of its peaks, in qualitative accordance with prediction (although rather more pronounced than expected).

According to quantum mechanical formalism the kinetic energy of the wave packets is given by the curvature of $\phi_k$. The shape parameter $A_k$ therefore determines the temperature of the system. It turns out that the temperature calculated in this manner is always too high if $A_k$ is allowed to vary between individual wave packets. Better agreement with experiment is obtained by the "semi-frozen" approximation, in which all $A_k$ are taken to be equal, changing in unison under the influence of a force that is averaged over all particles.

A more recent application of the WPD method is [KNAUP 99]. However, methodological progress for this technique is slow; see [HUBER 88, KOLAR 89, HERRERO 95, MARTINEZ 97].

# 7.4   Density Functional Molecular Dynamics (DFMD)

In a pioneering work Car and Parrinello introduced a method that permits a veritable dynamical simulation of quantum mechanical systems [CAR 85]. In the context of this "ab initio molecular dynamics" technique the only tribute to classical mechanics is the application of the Born-Oppenheimer approximation. The atomic cores (or "ions") consisting of the nucleus and the inner electronic shells are assumed to move according to classical laws, their masses being much larger than the single electron mass. But the valence and conduction electrons are represented by wave functions that are allowed to assume the configuration of least energy in the momentary field created by the ions (and by all other valence and conduction electrons). Let $\Psi_i(\mathbf{r})$ be the – mutually orthonormalized – wave functions of the $N$ electrons. The electron density at some position $\mathbf{r}$ is then given by

$$n(\mathbf{r}) \equiv \sum_{i=1}^{N} |\Psi_i(\mathbf{r})|^2 \qquad (7.65)$$

The momentary configuration of the (classical) ions is given by the set of ionic position vectors, $\{\mathbf{R}_l\}$. The ions produce a potential field $U(\mathbf{r}; \{\mathbf{R}_l\})$ which the electronic wave functions are invited to adjust to.

The energy of the system depends on the spatially varying electron density and on the ion potential $U(\ldots)$. To be exact, the expression for the total energy is

$$E(\{\Psi_i\}; \{\mathbf{R}_l\}) = E_1 + E_2 + E_3 + E_4 \qquad (7.66)$$

with

$$E_1 = \sum_{i=1}^{N} \int_V d\mathbf{r}\, \Psi_i^*(\mathbf{r}) \left[ -\frac{\hbar^2}{2m}\nabla^2 \right] \Psi_i(\mathbf{r}) \qquad (7.67)$$

$$E_2 = \int_V d\mathbf{r}\, U(\mathbf{r}; \{\mathbf{R}_l\})\, n(\mathbf{r}) \qquad (7.68)$$

$$E_3 = \frac{1}{2} \int_V \int_V d\mathbf{r}\, d\mathbf{r}' \frac{n(\mathbf{r})\, n(\mathbf{r}')}{|\mathbf{r} - \mathbf{r}'|} \qquad (7.69)$$

$$E_4 = E_{xc}[n(\mathbf{r})] \qquad (7.70)$$

$E_1$ gives the kinetic energy of the electrons, and their potential energy in the field created by the ions is given by $E_2$. The term $E_3$ accounts for the electrostatic interaction *between* the electrons. Finally, $E_{xc}$ stands for "exchange and correlation", representing the contribution of quantum mechanical exchange and correlation interactions to the total energy. There are various approximate expressions for this latter term. The most simple one, which has proved quite satisfactory in this context, is the so-called *local density approximation* (see [CAR 85]).

In practical work the wave functions $\Psi_i(\mathbf{r})$ are usually expanded in terms of plane waves,

$$\Psi_i(\mathbf{r}) = \sum_{\mathbf{k}} c_i(\mathbf{k}) e^{i\mathbf{k} \cdot \mathbf{r}} \qquad (7.71)$$

with up to several hundred terms per electron. The problem now is to find that set of expansion coefficients $\{c_i(\mathbf{k})\}$, i.e. those wave functions $\{\Psi_i\}$, which minimize the energy functional 7.66. Of course, the orthonormality condition

$$\int_V \Psi_i^*(\mathbf{r}, t)\, \Psi_j(\mathbf{r}, t)\, d\mathbf{r} = \delta_{ij} \tag{7.72}$$

must be met as well. Application of variational calculus to this problem leads to the so-called Kohn-Sham equations [KOHN 65] which may be solved by an iterative method. However, this procedure is too slow to permit a dynamical simulation.

Fortunately, we are already in possession of a powerful and efficient method for finding the minimum of a many-variable function: *simulated annealing*. The original formulation of this technique, as given by Kirkpatrick et al. [KIRKPATRICK 83], has been explained in the context of the statistical-mechanical Monte Carlo method (see Sec. 3.4.1). It may be employed here without alteration.

In addition, Car and Parrinello have suggested a variant of simulated annealing that is more in keeping with the spirit of dynamical simulation; they called their approach "dynamical simulated annealing":

Let $\mu$ denote an abstract (and at the moment arbitrary) "mass" assigned to each electronic wave function $\Psi_i$. We may then define an equally abstract "kinetic energy" pertaining to a temporal change of $\Psi_i$:

$$E_{kin} \equiv \frac{\mu}{2} \sum_{i=1}^{N} \int d\mathbf{r}\, \left|\dot{\Psi}_i\right|^2 \tag{7.73}$$

The formal analogy to mechanics is carried even further by the introduction of a Lagrangian

$$L = \sum_i \frac{\mu}{2} \int_V d\mathbf{r}\, \left|\dot{\Psi}_i(\mathbf{r})\right|^2 + \frac{M}{2}\sum_l \left|\dot{\mathbf{R}}_l\right|^2 - E(\{\Psi_i\}; \{\mathbf{R}_l\})$$

$$+ \sum_i \sum_j \lambda_{ij}\left[\int \Psi_i^*\Psi_j - \delta_{ij}\right] \tag{7.74}$$

Here $M$ is the ionic mass, and the Lagrange multipliers $\lambda_{ij}$ have been introduced to allow for the conditions 7.72. Application of the Lagrangian formalism of mechanics yields the "equations of motion"

$$\mu\ddot{\Psi}_i(\mathbf{r}, t) = -\frac{\delta E}{\delta \Psi_i^*(\mathbf{r}, t)} + \sum_j \lambda_{ij}\Psi_j(\mathbf{r}, t) \tag{7.75}$$

$$M\ddot{\mathbf{R}}_l = -\nabla_l E \tag{7.76}$$

Equation 7.76 describes the classical dynamics of the ions. The first equation, however, represents the abstract "motion" in the space of the electronic degrees of freedom. If we keep the "temperature" of this motion, as given by the "kinetic energy" $(\mu/2) \sum \left|\dot{\Psi}_i\right|^2$, small at all times, then the electronic subsystem will always remain close to the momentary minimum of the energy surface defined by the slowly varying ionic configuration.

To meet the requirement that the electronic degrees of freedom are to adjust quite fast to the varying energy landscape we have to choose the abstract mass $\mu$ rather small in comparison to the ionic masses. (A good choice is $\mu = 1.0$ atomic mass unit.)

If we were to leave the dynamic system 7.75-7.76 to its own devices, the electronic degrees of freedom would gradually assume the temperature of the ionic motion. To keep the temperature of the $\Psi_i$ small we may either rescale all $\dot{\Psi}_i$ from time to time or introduce one of the thermostats available from statistical-mechanical simulation; see Sec. 6.3.3.

The DFMD technique has become a major tool in computational material science. Current applications may be found by a web search, or in one of the regularly appearing survey articles, such as [VITEK 89, MAKRI 99, MAZZONE 99, OHNO 99].

# Chapter 8

# Hydrodynamics

*Fixed and co-moving grids*

The flow field $\mathbf{v}(\mathbf{r}, t)$ in a compressible viscous fluid obeys the equation of motion

$$\frac{\partial}{\partial t}\rho\mathbf{v} + \nabla \cdot [\rho\mathbf{v}\mathbf{v}] + \nabla p - \mu\nabla \cdot \mathbf{U} = 0 \qquad (8.1)$$

with $\mu$ denoting the viscosity, and the Navier-Stokes tensor $\mathbf{U}$ defined by

$$\mathbf{U} \equiv \nabla\mathbf{v} + (\nabla\mathbf{v})^T - \frac{2}{3}(\nabla \cdot \mathbf{v})\mathbf{I} \qquad (8.2)$$

(The coefficient in the last term is dependent on dimensionality; in two dimensions it is 1 instead of 2/3.)

This equation contains both advective (hyperbolic, that is) and diffusive (parabolic) terms. For small or vanishing viscosity the advective character is predominant, while in the viscous case the diffusive terms dominate. In the stationary case, i.e. for $\partial/\partial t = 0$, we are dealing with an elliptic equation.

The Navier-Stokes equation 8.1 is supplemented by the continuity equation for the mass,

$$\frac{\partial\rho}{\partial t} + \nabla \cdot \rho\mathbf{v} = 0 \qquad (8.3)$$

and by the equation for the conservation of energy,

$$\frac{\partial e}{\partial t} + \nabla \cdot [(e+p)\mathbf{v}] = 0 \tag{8.4}$$

where

$$e \equiv \rho\epsilon + \frac{\rho v^2}{2} \tag{8.5}$$

denotes the energy density ($\epsilon$ ... internal energy per unit mass of the fluid). Finally, an equation of state $p = p(\rho, \epsilon)$ coupling the pressure to density and thermal energy is required.

Equations 8.1-8.4 describe a perplexing multitude of phenomena, and it is advisable to stake out smaller sub-areas. If we make the viscosity negligible we find instead of 8.1 an equation describing the motion of an "ideal fluid" (Section 8.1). The air flow in the vicinity of an aircraft may be represented in this way. On the other hand, by taking into account the viscosity but neglecting the compressibility we arrive at equations that describe the flow of real liquids (Section 8.2).

Equation 8.1 does not contain the influence of gravity. If we add a term $\rho\mathbf{g}$ ($\mathbf{g}$ ... acceleration of gravity) the fluid will have a free surface capable of carrying waves. To calculate and visualize such phenomena one may use the MAC (*marker and cell*) method (see Section 8.2.3).

The partial differential equations 8.1-8.4, or their simplified versions, may be tackled using the techniques explained in Chapter 5. A quite different approach to numerical hydrodynamics has recently been suggested by the study of *lattice gas models* (see Section 8.3). These are a specific type of *cellular automata*, i.e. 2- or 3-dimensional bit patterns evolving according to certain rules.

Related to the lattice gas techniques is the slightly newer "Lattice Boltzmann method" of Section 8.3.2. Finally the "Direct Simulation Monte Carlo" technique introduced by Bird for the treatment of rarefied gas flow will be described in Section 8.4.

## 8.1   Compressible Flow without Viscosity

The frictionless flow of a fluid is described by the equations

$$\frac{\partial \rho}{\partial t} + \nabla \cdot \rho\mathbf{v} = 0 \tag{8.6}$$

$$\frac{\partial \rho\mathbf{v}}{\partial t} + \nabla \cdot [\rho\mathbf{v}\mathbf{v}] + \nabla p = 0 \tag{8.7}$$

$$\frac{\partial e}{\partial t} + \nabla \cdot [(e+p)\mathbf{v}] = 0 \tag{8.8}$$

In these *Eulerian* flow equations a laboratory-fixed coordinate system is assumed implicitely. The time derivative $\partial/\partial t$ is to be taken at a fixed point in space. However, the properties of a volume element that is moving along with the flowing substance will change according to the *Lagrange derivative*

$$\frac{d}{dt} \equiv \frac{\partial}{\partial t} + \mathbf{v} \cdot \nabla \tag{8.9}$$

so that the above equations may alternatively be written in the Lagrange form

$$\frac{d\rho}{dt} = -\rho \nabla \cdot \mathbf{v} \tag{8.10}$$

$$\rho \frac{d\mathbf{v}}{dt} = -\nabla p \tag{8.11}$$

$$\frac{de}{dt} = -(e+p)\nabla p - (\mathbf{v} \cdot \nabla) p$$

$$= -e(\nabla \cdot \mathbf{v}) - \nabla \cdot (p\mathbf{v}) \tag{8.12}$$

Using 8.5 the last equation may be cast into the form

$$\frac{d\epsilon}{dt} = -\frac{p}{\rho} (\nabla \cdot \mathbf{v}) \tag{8.13}$$

## 8.1.1    Explicit Eulerian Methods

Euler's equations 8.6-8.8 may always be written in the standard conservative-advective form that has been discussed at the beginning of Chapter 5:

$$\frac{\partial \mathbf{u}}{\partial t} = -\frac{\partial \mathbf{j}_x}{\partial x} - \frac{\partial \mathbf{j}_y}{\partial y} - \frac{\partial \mathbf{j}_z}{\partial z} \tag{8.14}$$

with

$$\mathbf{u} = \begin{pmatrix} \rho \\ \rho v_x \\ \rho v_y \\ \rho v_z \\ e \end{pmatrix}, \quad \mathbf{j}_x = \begin{pmatrix} \rho v_x \\ \rho v_x^2 + p \\ \rho v_y v_x \\ \rho v_z v_x \\ (e+p)v_x \end{pmatrix}, \quad \mathbf{j}_y = \begin{pmatrix} \rho v_y \\ \rho v_x v_y \\ \rho v_y^2 + p \\ \rho v_z v_y \\ (e+p)v_y \end{pmatrix}, \quad \mathbf{j}_z = \begin{pmatrix} \rho v_z \\ \rho v_x v_z \\ \rho v_y v_z \\ \rho v_z^2 + p \\ (e+p)v_z \end{pmatrix}$$

$$\tag{8.15}$$

Therefore the entire arsenal of methods given in Chapter 5 for the numerical treatment of conservative-advective equations – Lax, Lax-Wendroff, leapfrog – may be invoked to solve equ. 8.14. As a simple example let us write down the Lax algorithm for the one-dimensional case (see [POTTER 80]):

**Explicit Euler / Lax:**

$$\rho_j^{n+1} = \frac{1}{2}\left(\rho_{j+1}^n + \rho_{j-1}^n\right)$$

$$-\frac{\Delta t}{2\Delta x}\left(\rho_{j+1}^n v_{j+1}^n - \rho_{j-1}^n v_{j-1}^n\right) \tag{8.16}$$

$$\rho_j^{n+1} v_j^{n+1} = \frac{1}{2}\left(\rho_{j+1}^n v_{j+1}^n + \rho_{j-1}^n v_{j-1}^n\right)$$

$$-\frac{\Delta t}{2\Delta x}\left[\rho_{j+1}^n (v_{j+1}^n)^2 + p_{j+1}^n - \rho_{j-1}^n (v_{j-1}^n)^2 - p_{j-1}^n\right] \tag{8.17}$$

$$e_j^{n+1} = \frac{1}{2}\left(e_{j+1}^n - e_{j-1}^n\right)$$

$$-\frac{\Delta t}{2\Delta x}\left[\left(e_{j+1}^n + p_{j+1}^n\right)v_{j+1}^n - \left(e_{j-1}^n + p_{j-1}^n\right)v_{j-1}^n\right] \tag{8.18}$$

## 8.1.2    Particle-in-Cell Method (PIC)

For simplicity we will here consider an ideal gas. Also, at this level we want to avoid having to deal with the effects of thermal conductivity. Our assumption therefore is that the gas flows so fast that the adiabatic equation of state holds. In a moving mass element of the fluid, then, the quotient $p/\rho^\gamma = c$ will be constant – in other words, its Lagrangian time derivative is zero:

$$\frac{\partial c}{\partial t} + \mathbf{v} \cdot \nabla c = 0 \tag{8.19}$$

Together with 8.6 this yields a continuity equation for the quantity $\rho c$,

$$\frac{\partial}{\partial t}[\rho c] + \nabla \cdot [\rho c \mathbf{v}] = 0 \tag{8.20}$$

Thus the equations for the inviscid flow of an ideal gas that are to be treated by the PIC- (*particle in cell*-) method read

$$\frac{\partial \rho}{\partial t} + \nabla \cdot (\rho \mathbf{v}) = 0 \tag{8.21}$$

$$\frac{\partial \rho \mathbf{v}}{\partial t} + \nabla \cdot (\rho \mathbf{v} \mathbf{v}) = -\nabla p \tag{8.22}$$

$$\frac{\partial}{\partial t}(\rho c) + \nabla \cdot (\rho c \mathbf{v}) = 0 \tag{8.23}$$

With no harm to generality we may consider the two-dimensional case. First we discretize the spatial axes to obtain an *Eulerian* (lab-fixed) lattice of cells with side lengths $\Delta x = \Delta y = \Delta l$. A representation of the local density is achieved by filling each cell with a variable number of particles; to keep statistical density fluctuations low the number of particles in a cell should not be too small. The particles are not meant to represent atoms or molecules but "fluid elements" whose properties at time $t_n$ are described by the vectors

$$\mathbf{u}_k^n \equiv \{\mathbf{r}_k^n, \mathbf{v}_k^n, c_k^n\} \qquad k = 1, \dots N \tag{8.24}$$

The net properties of the Eulerian cells are then simply sums over the particles they contain:

$$\rho_{i,j}^n = \frac{m}{(\Delta l)^2} \sum_{k=1}^{N} \delta\left[\mathbf{r}_k^n(i,j)\right] \tag{8.25}$$

$$(\rho \mathbf{v})_{i,j}^n = \frac{m}{(\Delta l)^2} \sum_{k=1}^{N} \mathbf{v}_k^n \delta\left[\mathbf{r}_k^n(i,j)\right] \tag{8.26}$$

$$(\rho c)_{i,j}^n = \frac{m}{(\Delta l)^2} \sum_{k=1}^{N} c_k^n \delta\left[\mathbf{r}_k^n(i,j)\right] \tag{8.27}$$

where we have used the short notation

$$\delta\left[\mathbf{r}(i,j)\right] \equiv \delta\left[\mathrm{int}(\frac{x}{\Delta l}) - i\right] \delta\left[\mathrm{int}(\frac{y}{\Delta l}) - j\right] \tag{8.28}$$

To update the cell velocities – the velocities *within the cells*, that is – we write 8.22 in the form

$$\rho \frac{\partial \mathbf{v}}{\partial t} = -\nabla p - \mathbf{v} \frac{\partial \rho}{\partial t} - \nabla \cdot (\rho \mathbf{v} \mathbf{v}) \tag{8.29}$$

and for a moment neglect the last two terms on the right hand side. The remaining equation describes the effect of the pressure gradient on the cell velocities. By discretizing and using the equation of state to evaluate the cell pressures $p_{i,j}$ we obtain the preliminary new values

$$v_{x,i,j}^{n+1} = v_{x,i,j}^n - a\left(p_{i+1,j}^n - p_{i-1,j}^n\right) \tag{8.30}$$

$$v_{y,i,j}^{n+1} = v_{y,i,j}^n - a\left(p_{i,j+1}^n - p_{i,j-1}^n\right) \tag{8.31}$$

with

$$a \equiv \Delta t \, / 2(\Delta l) \rho_{i,j}^n \tag{8.32}$$

Each particle (fluid element) $k$ may now be given a new value of the velocity and of the quantity $c$. We assume that the particles simply adopt the properties pertaining to the Euler cell they inhabit (local equilibrium), writing

$$\mathbf{v}_k^{n+1} = \mathbf{v}_{i,j}^{n+1} \quad \text{and} \quad c_k^{n+1} = p_{i,j}^n / (\rho_{i,j}^n)^\gamma \tag{8.33}$$

Now we attend to the Lagrangian transport terms in equation 8.29. The simplest way to account for their effect is to let the fluid particles move along with suitable velocities. Defining the time centered cell velocities

$$\mathbf{v}_{i,j}^{n+1/2} = \frac{1}{2}\left[\mathbf{v}_{i,j}^{n+1} + \mathbf{v}_{i,j}^n\right] \tag{8.34}$$

we compute the particle velocities by taking a weighted sum over the adjacent Eulerian cells:

$$v_k^{n+1/2} = \frac{1}{(\Delta l)^2} \sum_{(ij)} a_{(ij)} v_{(ij)}^{n+1/2} \tag{8.35}$$

The weights $a_{(ij)}$ are the overlap areas of a square of side length $\Delta l$ centered around particle $k$ and the nearest Euler cells $(ij)$. (We have encountered this kind of area weighting before, in conjunction with the *particle-mesh* method of 6.5.2; see Fig. 6.9.) With the updated positions

$$r_k^{n+1} = r_k^n + \Delta t \, v_k^{n+1/2} \tag{8.36}$$

and the quantities of equ. 8.33 we have completed the new vector of particle properties, $u_k^{n+1}$. A step-by-step description of the PIC method is given in Figure 8.1.

### 8.1.3   Smoothed Particle Hydrodynamics (SPH)

The PIC technique is a cross-breed between an Eulerian and a Lagrangian method. The velocity change due to pressure gradients is computed using a fixed grid of Euler cells, but the transport of momentum and energy is treated à la Lagrange, namely by letting the fluid elements (particles) move in continuous space. The rationale for switching back and forth between the two representations is that equation 8.29 involves a pressure gradient. Tradition has it that gradients are most easily evaluated on a regular grid – see 8.30-8.31. In contrast, the transport of conserved quantities is simulated quite naturally using a particle picture – see 8.36.

However, there have been very fruitful attempts to altogether avoid the use of the Euler lattice. All information about the state of the moving fluid is contained in the vectors $u_k$ ($k = 1, \ldots N$), and steps 1 and 2 in the PIC method (Fig. 8.1) are really just a methodological detour through Euler territory, with the sole purpose of evaluating density and pressure and differencing the latter. In principle, it should be possible to determine the pressure gradients, and thus the forces acting on the fluid elements, without ever leaving the particle picture.

If we abandon Euler cells we have to provide for some consistent representation of the spatially continuous fluid density. In the PIC method the average density within a cell was determined by the number of point particles in that cell. Lucy [LUCY 77] and Gingold and Monaghan [GINGOLD 77, MONAGHAN 92] pointed out that by loading each particle with a spatially extended interpolation kernel one may define an average density at any point in space as a sum over the individual contributions. Let $w(r - r_i)$ denote the interpolation kernel centered around the position of particle $i$; the estimated density at $r$ is then

$$\langle \rho(r) \rangle = \sum_{i=1}^{N} m_i \, w(r - r_i) \tag{8.45}$$

**PIC method (2-dimensional):** At time $t_n$ the state of the fluid is represented by $N$ particles with the property vectors $\mathbf{u}_k^n \equiv \{\mathbf{r}_k^n, \mathbf{v}_k^n, c_k^n\}$ ($k = 1, \ldots N$). In each Eulerian cell of side length $\Delta l$ there should be at least $\approx 100$ particles.

1. Compute, for each Euler cell $(i, j)$, the cell properties

$$\rho_{i,j}^n = \frac{m}{(\Delta l)^2} \sum_{k=1}^{N} \delta \left[ \mathbf{r}_k^n(i, j) \right] \tag{8.37}$$

$$(\rho \mathbf{v})_{i,j}^n = \frac{m}{(\Delta l)^2} \sum_{k=1}^{N} \mathbf{v}_k^n \delta \left[ \mathbf{r}_k^n(i, j) \right] \tag{8.38}$$

$$(\rho c)_{i,j}^n = \frac{m}{(\Delta l)^2} \sum_{k=1}^{N} c_k^n \delta \left[ \mathbf{r}_k^n(i, j) \right] \tag{8.39}$$

2. Using the equation of state to evaluate cell pressures $p_{i,j}$, compute new (preliminary) flow velocities according to

$$v_{x,i,j}^{n+1} = v_{x,i,j}^n - a \left( p_{i+1,j}^n - p_{i-1,j}^n \right) \tag{8.40}$$

$$v_{y,i,j}^{n+1} = v_{y,i,j}^n - a \left( p_{i,j+1}^n - p_{i,j-1}^n \right) \tag{8.41}$$

with $a \equiv \Delta t \big/ 2(\Delta l) \rho_{i,j}^n$. For each fluid particle $k$ we now have $\mathbf{v}_k^{n+1} = \mathbf{v}_{i,j}^{n+1}$ and $c_k^{n+1} = p_{i,j}^n / (\rho_{i,j}^n)^\gamma$.

3. From the time-centered cell velocities $\mathbf{v}_{i,j}^{n+1/2} \equiv \left[ \mathbf{v}_{i,j}^{n+1} + \mathbf{v}_{i,j}^n \right] \big/ 2$ compute for each particle $k$ an intermediate velocity

$$\mathbf{v}_k^{n+1/2} = \frac{1}{(\Delta l)^2} \sum_{(ij)} a_{(ij)} \mathbf{v}_{(ij)}^{n+1/2} \tag{8.42}$$

using suitable weights $a_{(ij)}$ (see text); calculate new particle positions

$$\mathbf{r}_k^{n+1} = \mathbf{r}_k^n + \Delta t \, \mathbf{v}_k^{n+1/2} \tag{8.43}$$

4. Each particle is now given the new state vector

$$\mathbf{u}_k^{n+1} \equiv \{\mathbf{r}_k^{n+1}, \mathbf{v}_k^{n+1}, c_k^{n+1}\} \tag{8.44}$$

Figure 8.1: Particle-in-cell method

where $m_i$ is the mass of the particle (i.e. the fluid element). More generally, *any* spatially varying property $A(\mathbf{r})$ of the fluid may be represented by its "smoothed particle estimate"

$$\langle A(\mathbf{r})\rangle = \sum_{i=1}^{N} m_i \frac{A(\mathbf{r}_i)}{\rho(\mathbf{r}_i)}\, w(\mathbf{r} - \mathbf{r}_i) \tag{8.46}$$

(where $\rho(\mathbf{r}_i)$ now denotes the average 8.45 taken at the position $\mathbf{r}_i$). The function $w(\mathbf{s})$, which by the various authors has been called smoothing, broadening, weighting or interpolating kernel, is most conveniently assumed to be a Gaussian. In three dimensions, then,

$$w(\mathbf{s}) = \frac{1}{\pi^{3/2} d^3}\, e^{-s^2/d^2} \tag{8.47}$$

with an arbitrary width $d$. If the width is small, the density interpolant will fluctuate rather heavily; if it is too large, the summations cannot be restricted to nearby particles and thus become time-consuming. In practice one chooses $d$ such that the average number of neighboring particles spanned by the Gaussian is about 5 for two dimensions and 15 in the three-dimensional case. Other functional forms than the Gaussian are possible and sometimes even lead to better results.

We return now to the Lagrangian equations of motion for mass, momentum and energy, equs. 8.10, 8.11 and 8.13, and try to rewrite them consistently in smoothed particle form. In keeping with the somewhat intuitive character of the SPH method, various ways of defining the quantity $A(\mathbf{r})$ to be interpolated according to 8.46 have been tried out. For instance, in the momentum equation $d\mathbf{v}/dt = -\nabla p/\rho$ one might interpolate $\rho$ and $\nabla p$ directly, inserting the results on the right hand side. It turns out that this procedure would not conserve linear and angular momentum [MONAGHAN 92]. Instead, one uses the identity

$$\frac{1}{\rho}\nabla p = \nabla\left(\frac{p}{\rho}\right) + \frac{p}{\rho^2}\nabla\rho \tag{8.48}$$

and the SPH expressions for $A \equiv p/\rho$ and $A \equiv \rho$ to write the velocity equation as

$$\frac{d\mathbf{v}_i}{dt} = -\sum_{k=1}^{N} m_k \left(\frac{p_k}{\rho_k^2} + \frac{p_i}{\rho_i^2}\right) \nabla_i w_{ik} \tag{8.49}$$

with $w_{ik} \equiv w(\mathbf{r}_{ik}) \equiv w(\mathbf{r}_k - \mathbf{r}_i)$. If $w_{ik}$ is Gaussian, this equation describes the motion of particle $i$ under the influence of central pair forces

$$\mathbf{F}_{ik} = -m_i m_k \left(\frac{p_k}{\rho_k^2} + \frac{p_i}{\rho_i^2}\right) \frac{2\mathbf{r}_{ik}}{d^2} w_{ik} \tag{8.50}$$

Similar considerations lead to the SPH equivalents of the other Lagrangian flow equations,

$$\frac{d\rho_i}{dt} = \sum_{k=1}^{N} m_k\, \mathbf{v}_{ik} \cdot \nabla_i w_{ik} \tag{8.51}$$

where $\mathbf{v}_{ik} \equiv \mathbf{v}_k - \mathbf{v}_i$, and

$$\frac{d\epsilon_i}{dt} = -\sum_{k=1}^{N} m_k \left( \frac{p_k}{\rho_k^2} + \frac{p_i}{\rho_i^2} \right) \mathbf{v}_{ik} \cdot \nabla_i w_{ik} \qquad (8.52)$$

The equation of motion for the density $\rho$ need not be integrated. Instead, equ. 8.45 may be invoked to find and estimate for the density at $\mathbf{r}_i$ once all particle positions are known. Note that in addition to $\rho_i$, $\mathbf{v}_i$ and $\epsilon_i$ the position $\mathbf{r}_i$ must also be updated to complete a time step cycle. The obvious relation

$$\frac{d\mathbf{r}_i}{dt} = \mathbf{v}_i \qquad (8.53)$$

may be used, but Monaghan has shown that the less obvious formula

$$\frac{d\mathbf{r}_i}{dt} = \mathbf{v}_i + \sum_{k=1}^{N} \frac{m_k}{\bar{\rho}_{ik}} \mathbf{v}_{ik} \, w_{ik} \qquad (8.54)$$

with $\bar{\rho}_{ik} \equiv (\rho_i + \rho_k)/2$ leaves angular and linear momentum conservation intact while offering the advantage that nearby particles will have similar velocities [MONAGHAN 89].

Equations 8.51, 8.49, 8.53 and 8.52 may be solved simultaneously by some suitable algorithm (see Chapter 4). The leapfrog algorithm has often been applied, but the use of predictor-corrector and Runge-Kutta schemes has also been reported. One out of many possible integration procedures is the following variant of the half-step technique ([MONAGHAN 89]):

Given all particle positions at time $t_n$, the local density at $\mathbf{r}_i$ is computed from the interpolation formula 8.45. Writing equs. 8.49 and 8.52 as

$$\frac{d\mathbf{v}_i}{dt} = \mathbf{F}_i \quad \text{and} \quad \frac{d\epsilon_i}{dt} = Q_i \qquad (8.55)$$

the predictors

$$\mathbf{v}_i^{n+1} = \mathbf{v}_i^n + \Delta t \, \mathbf{F}_i^n, \quad \epsilon_i^{n+1} = \epsilon_i^n + \Delta t \, Q_i^n \qquad (8.56)$$

and

$$\mathbf{r}_i^{n+1} = \mathbf{r}_i^n + \Delta t \, \mathbf{v}_i^n \qquad (8.57)$$

are calculated. Mid-point values of $\mathbf{r}_i$, $\mathbf{v}_i$ and $\epsilon_i$ are determined according to

$$\mathbf{r}_i^{n+1/2} = \left( \mathbf{r}_i^n + \mathbf{r}_i^{n+1} \right)/2 \qquad (8.58)$$

etc. From these, mid-point values of $\rho_i$, $\mathbf{F}_i$ and $Q_i$ are computed and inserted in correctors of the type

$$\mathbf{v}_i^{n+1} = \mathbf{v}_i^n + \Delta t \, \mathbf{F}_i^{n+1/2} \qquad (8.59)$$

Note that the equation of motion for the density, equ. 8.51, is not integrated numerically. Using the interpolation formula for $\rho$ takes somewhat longer, since

**Smoothed particle hydrodynamics:** At time $t_n$ the state of the fluid is represented by $N$ particles with masses $m_i$ and the property vectors $\mathbf{u}_i^n \equiv \{\mathbf{r}_i^n, \mathbf{v}_i^n, \epsilon_i^n\}$ $(i = 1, \ldots N)$. (In the case of an ideal gas undergoing adiabatic flow, the specific energy $\epsilon$ may be replaced by $c \equiv p/\rho^\gamma = \epsilon(\gamma - 1)/\rho^{\gamma - 1}$). A suitable interpolation kernel is assumed, e.g. $w(\mathbf{s}) = (1/\pi^{3/2}d^3)exp\{-s^2/d^2\}$, with the width $d$ chosen so as to span about 5 (in 2 dimensions) or 15 (3-d) neighbors.

1. At each particle position $\mathbf{r}_i$ the density $\rho_i$ is computed by interpolation:

$$\rho_i = \sum_{k=1}^{N} m_k \, w(\mathbf{r}_{ik}) \qquad (8.60)$$

2. From the given equation of state $p = p(\rho, \epsilon)$ compute the pressures $p_i = p(\rho_i, \epsilon_i)$.

3. Integrate the equations of motion

$$\frac{d\mathbf{r}_i}{dt} = \mathbf{v}_i \quad (\text{or} \quad \text{equ. 8.54}) \qquad (8.61)$$

$$\frac{d\mathbf{v}_i}{dt} = -\sum_{k=1}^{N} m_k \left( \frac{p_k}{\rho_k^2} + \frac{p_i}{\rho_i^2} \right) \nabla_i w_{ik} \qquad (8.62)$$

$$\frac{d\epsilon_i}{dt} = -\sum_{k=1}^{N} m_k \left( \frac{p_k}{\rho_k^2} + \frac{p_i}{\rho_i^2} \right) \mathbf{v}_{ik} \cdot \nabla_i w_{ik} \qquad (8.63)$$

over one time step by some suitable integrator (Runge-Kutta, or the simple procedure 8.55-8.59) to obtain

$$\mathbf{u}_i^{n+1} \equiv \{\mathbf{r}_i^{n+1}, \mathbf{v}_i^{n+1}, \epsilon_i^{n+1}\} \qquad i = 1, \ldots N \qquad (8.64)$$

A modification of this scheme is obtained by including the density $\rho_i$ in the state vector of particle $i$ and integrating the pertinent equation of motion, 8.51. The time step integrations for $\mathbf{r}$, $\mathbf{v}$, $\rho$ and $\epsilon$ may then be performed simultaneously, and the evaluation of the density according to 8.60 is omitted. This procedure works faster, but exact mass conservation is not guaranteed.

Figure 8.2: Smoothed particle hydrodynamics (SPH)

the summation in 8.45 has to be performed separately, but mass conservation is better fulfilled than by integrating 8.51.

Figure 8.2 gives an overview of one time step in a basic version of the SPH procedure.

It should be noted that the SPH technique, although it is here discussed in conjunction with compressible inviscid flow, may be applied to other flow problems as well. Incompressibility may be handled by using an equation of state that keeps compressibility effects below a few percent [MONAGHAN 92], and the influence of viscosity is best accounted for by an additional term in the equations of motions for momentum and energy, equs. 8.49 and 8.52, thus:

$$\frac{d\mathbf{v}_i}{dt} = -\sum_{k=1}^{N} m_k \left( \frac{p_k}{\rho_k^2} + \frac{p_i}{\rho_i^2} + \Pi_{ik} \right) \nabla_i w_{ik} \tag{8.65}$$

$$\frac{d\epsilon_i}{dt} = -\sum_{k=1}^{N} m_k \left( \frac{p_k}{\rho_k^2} + \frac{p_i}{\rho_i^2} + \Pi_{ik} \right) \mathbf{v}_{ik} \cdot \nabla_i w_{ik} \tag{8.66}$$

The artificial viscosity term $\Pi_{ik}$ is modeled in the following way:

$$\Pi_{ik} = \begin{cases} \dfrac{-\alpha \bar{c}_{ik} \mu_{ik} + \beta \mu_{ik}^2}{\bar{\rho}_{ik}} & \text{if } \mathbf{v}_{ik} \cdot \mathbf{r}_{ik} < 0 \\ 0 & \text{if } \mathbf{v}_{ik} \cdot \mathbf{r}_{ik} \geq 0 \end{cases} \tag{8.67}$$

with $c$ denoting the speed of sound, $\mu$ defined by

$$\mu_{ik} = \frac{(\mathbf{v}_{ik} \cdot \mathbf{r}_{ik}) d}{r_{ik}^2 + \eta^2} \tag{8.68}$$

and the conventions $a_{ik} \equiv a_k - a_i$ and $\bar{a}_{ik} \equiv (a_i + a_k)/2$. This form of $\Pi$ takes care of the effects of shear and bulk viscosity. The parameters $\alpha$ and $\beta$ are not critical, but should be near $\alpha = 1$ and $\beta = 2$ for best results [MONAGHAN 92]. The quantity $\eta$ prevents singularities for $r_{ik} \approx 0$. It should be chosen according to $\eta^2 = 0.01d^2$.

Another physical feature that has been excluded from our discussion but may be treated in the framework of SPH is thermal conduction. A suitable term representing the exchange of thermal energy between particles is given in [MONAGHAN 89].

A long-standing problem in the application of SPH has been the treatment of interfaces. The spatial region beyond a boundary has to be populated with particles that interact with the particles representing the moving fluid. The properties – in particular the velocities – of these dummy particles have to be chosen with care such that either a free surface or a "sticky" solid boundary is represented in a consistent manner.

Recently, both the free surface and the no-slip boundary problem have been tackled successfully. Nugent and Posch [NUGENT 00] devised a method to treat free surfaces, and Ivanov [IVANOV 00] found a way to represent rough interfaces in the smoothed particle picture.

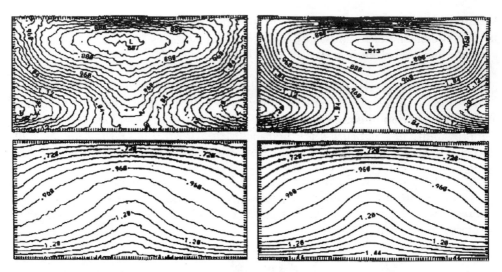

Figure 8.3: Comparison of Smoothed Particle Hydrodynamics with an Eulerian finite-difference calculation. The density (above) and temperature (below) contours for a stationary Rayleigh-Bénard flow are shown. Left: SPH; right: Euler. (From [HOOVER 99], with kind permission by the author)

A standard test exercise in numerical hydrodynamics is the simulation of "Rayleigh-Bénard" convection: a slab of fluid that is heated from below and cooled from above may form stable convective rolls which transport heat from the bottom to the top. Figure 8.3 compares the performance of Smoothed Particle Hydrodynamics with that of an Euler-type calculation.[HOOVER 99] Computing times are comparable for both calculations, and the results are in good agreement. There are, of course, no fluctuations in the continuum calculation, in contrast to the particle simulation. However, the SPH program is extremely simple, being just a MD program with a special kind of interaction, while the effort that goes into developing a Euler finite-difference scheme is considerable.

## 8.2 Incompressible Flow with Viscosity

Assuming $d\rho/dt = 0$ in 8.3 we find

$$\nabla \cdot \mathbf{v} = 0 \tag{8.69}$$

The flow of an incompressible liquid is necessarily source-free. Furthermore, 8.69 implies that

$$\nabla \cdot (\nabla \mathbf{v}) + \nabla \cdot (\nabla \mathbf{v})^T = \nabla^2 \mathbf{v} \tag{8.70}$$

so that the Navier-Stokes equation now assumes the form

$$\frac{\partial \mathbf{v}}{\partial t} + (\mathbf{v} \cdot \nabla)\mathbf{v} = -\nabla \bar{p} + \nu \nabla^2 \mathbf{v} \tag{8.71}$$

with $\nu \equiv \mu/\rho$ and $\bar{p} \equiv p/\rho$.

The two classic techniques for the numerical treatment of 8.69 and 8.71 are the *vorticity* and the *pressure* method.

## 8.2.1 Vorticity Method

Taking the rotation of equ. 8.71 we obtain

$$\frac{\partial \mathbf{w}}{\partial t} + (\mathbf{v} \cdot \nabla)\mathbf{w} = \nu \nabla^2 \mathbf{w} \tag{8.72}$$

where we have introduced the *vorticity* $\mathbf{w} \equiv \nabla \times \mathbf{v}$. We can see that the vorticity is transported both by an advective process $(\mathbf{v} \cdot \nabla \mathbf{w})$ and by viscous diffusion.

Since the velocity has no divergence it may be written as the rotation of a *streaming function* $\mathbf{u}$. The definition

$$\mathbf{v} \equiv \nabla \times \mathbf{u} \tag{8.73}$$

does not determine the function $\mathbf{u}$ uniquely; we are free to require that $\nabla \cdot \mathbf{u} = 0$. Thus the relations that provide the starting point for the vorticity method read

$$\frac{\partial \mathbf{w}}{\partial t} + (\mathbf{v} \cdot \nabla)\mathbf{w} = \nu \nabla^2 \mathbf{w} \tag{8.74}$$

$$\nabla^2 \mathbf{u} = -\mathbf{w} \tag{8.75}$$

$$\mathbf{v} = \nabla \times \mathbf{u} \tag{8.76}$$

In the two-dimensional case the vectors $\mathbf{u}$ and $\mathbf{w}$ have only $z$-components and may be treated as pseudoscalars:

$$\frac{\partial w}{\partial t} = \nu \nabla^2 w - (v_x \partial_y - v_y \partial_x) w \tag{8.77}$$

$$\nabla^2 u = -w \tag{8.78}$$

$$\mathbf{v} = u \nabla \times \mathbf{e}_z = \begin{pmatrix} \partial_y u \\ -\partial_x u \end{pmatrix} \tag{8.79}$$

The proven numerical method for solving these equations, a modification of the Lax-Wendroff scheme, is described in Figure 8.4. The stability of the method is once more governed by the CFL condition (see Section 5.1),

$$\Delta t \leq \frac{2\Delta l}{\sqrt{2} v_{max}} \tag{8.80}$$

In addition, the presence of diffusive terms implies the restriction

$$\Delta t \leq \frac{(\Delta l)^2}{\nu} \tag{8.81}$$

**Vorticity method (2-dimensional):**

Let the flow field at time $t_n$ be given by $u_{i,j}^n$ and $w_{i,j}^n$. For simplicity, let $\Delta y = \Delta x \equiv \Delta l$.

1. Auxiliary quantities:

$$v_{x,i,j+1}^n = \frac{1}{2\Delta l}\left(u_{i,j+2}^n - u_{i,j}^n\right) \tag{8.82}$$

$$v_{y,i,j+1}^n = -\frac{1}{2\Delta l}\left(u_{i+1,j+1}^n - u_{i-1,j+1}^n\right) \tag{8.83}$$

$$
\begin{aligned}
w_{i,j+1}^{n+1/2} = {} & \frac{1}{4}\left[w_{i,j}^n + w_{i+1,j+1}^n + w_{i,j+2}^n + w_{i-1,j+1}^n\right] \\
& -\frac{\Delta t}{4\Delta l}v_{x,i,j+1}^n\left(w_{i+1,j+1}^n - w_{i-1,j+1}^n\right) \\
& -\frac{\Delta t}{4\Delta l}v_{y,i,j+1}^n\left(w_{i,j+2}^n - w_{i,j}^n\right)
\end{aligned}
\tag{8.84}
$$

   etc., for the 4 lattice points nearest to $(i,j)$. Thus the viscous terms are being neglected for the time being (compare 8.74).

2. From the Poisson equation 8.75 the streaming function is also determined at half-step time, using diagonal differencing (see Section 1.3):

$$u_{i,j+1}^{n+1/2} + u_{i,j-1}^{n+1/2} + u_{i+2,j+1}^{n+1/2} + u_{i+2,j-1}^{n+1/2} - 4u_{i+1,j}^{n+1/2} = -w_{i+1,j}^{n+1/2}(\Delta l)^2 \tag{8.85}$$

3. Now follows the integration step proper, the viscous term included:

$$v_{x,i,j}^{n+1/2} = \frac{1}{2\Delta l}\left(u_{i,j+1}^{n+1/2} - u_{i,j-1}^{n+1/2}\right) \tag{8.86}$$

$$v_{y,i,j}^{n+1/2} = -\frac{1}{2\Delta l}\left(u_{i+1,j}^{n+1/2} - u_{i-1,j}^{n+1/2}\right) \tag{8.87}$$

$$
\begin{aligned}
w_{i,j}^{n+1} = {} & w_{i,j}^n - \frac{\Delta t}{2\Delta l}v_{x,i,j}^{n+1/2}\left(w_{i+1,j}^{n+1/2} - w_{i-1,j}^{n+1/2}\right) \\
& -\frac{\Delta t}{2\Delta l}v_{y,i,j}^{n+1/2}\left(w_{i,j+1}^{n+1/2} - w_{i,j-1}^{n+1/2}\right) \\
& +\frac{\nu\Delta t}{2(\Delta l)^2}\left(w_{i+1,j-1}^n + w_{i-1,j-1}^n + w_{i-1,j+1}^n + w_{i+1,j+1}^n - 4w_{i,j}^n\right)
\end{aligned}
\tag{8.88}
$$

Figure 8.4: Vorticity method

## 8.2.2    Pressure Method

Going back to the Navier-Stokes equation for incompressible flow, we now take the divergence (instead of rotation) of 8.71 and use the identity

$$\nabla \cdot (\mathbf{v} \cdot \nabla)\mathbf{v} = (\nabla \mathbf{v}) : (\nabla \mathbf{v}) \tag{8.89}$$

(with $\mathbf{A} : \mathbf{B} \equiv \sum_i \sum_j A_{ij} B_{ji}$) to obtain the set of equations

$$\frac{\partial \mathbf{v}}{\partial t} + (\mathbf{v} \cdot \nabla)\mathbf{v} = -\nabla \bar{p} + \nu \nabla^2 \mathbf{v} \tag{8.90}$$

$$\nabla^2 \bar{p} = -(\nabla \mathbf{v}) : (\nabla \mathbf{v}) \tag{8.91}$$

which provide the basis for the pressure method.

In the two-dimensional case these equations read

$$\frac{\partial v_x}{\partial t} = -\frac{\partial \bar{p}}{\partial x} + \nu \left[ \frac{\partial^2 v_x}{\partial x^2} + \frac{\partial^2 v_x}{\partial y^2} \right] - \frac{\partial v_x^2}{\partial x} - \frac{\partial v_x v_y}{\partial y} \tag{8.92}$$

$$\frac{\partial v_y}{\partial t} = -\frac{\partial \bar{p}}{\partial y} + \nu \left[ \frac{\partial^2 v_y}{\partial x^2} + \frac{\partial^2 v_y}{\partial y^2} \right] - \frac{\partial v_y^2}{\partial y} - \frac{\partial v_x v_y}{\partial x} \tag{8.93}$$

$$\frac{\partial^2 \bar{p}}{\partial x^2} + \frac{\partial^2 \bar{p}}{\partial y^2} = -\left[ \left( \frac{\partial v_x}{\partial x} \right)^2 + 2 \left( \frac{\partial v_x}{\partial y} \right) \left( \frac{\partial v_y}{\partial x} \right) + \left( \frac{\partial v_y}{\partial y} \right)^2 \right] \tag{8.94}$$

When attempting to solve these equations by a finite difference scheme we have to make sure that the divergence condition $\nabla \cdot \mathbf{v} = 0$ will stay intact in the course of the calculation. To achieve this, Harlow and Welch have suggested the following kind of discretisation ([HARLOW 65], see also [POTTER 80]):

The grid values of the pressure $p_{i,j}$ are taken to be localized at the centers of the Euler cells, while the velocity components $v_{x,i,j}$ and $v_{y,i,j}$ are placed at the right and upper box sides, respectively (see Fig. 8.5). The divergence of the velocity is then approximated by

$$D_{i,j} \equiv \frac{1}{\Delta l} [v_{x,i,j} - v_{x,i-1,j}] + \frac{1}{\Delta l} [v_{y,i,j} - v_{y,i,j-1}] \tag{8.95}$$

or, in "geographical" notation,

$$D_C \equiv \frac{1}{\Delta l} [v_{x,C} - v_{x,W}] + \frac{1}{\Delta l} [v_{y,C} - v_{y,S}] \tag{8.96}$$

The requirement of vanishing divergence then reads simply $D_C = 0$.

Using this staggered grid, the Navier-Stokes equations 8.92-8.93 are now treated à la Lax (all terms on the right hand side having the time index $n$):

$$v_{x,C}^{n+1} = \frac{1}{4} [v_{x,N} + v_{x,E} + v_{x,S} + v_{x,W}] - \frac{\Delta t}{2\Delta l} [v_{x,E}^2 - v_{x,W}^2]$$

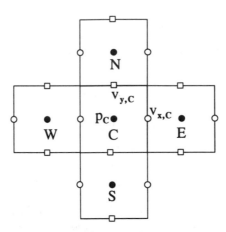

Figure 8.5: Grid structure in the pressure method

$$-\frac{\Delta t}{2\Delta l}\left[\frac{1}{2}\left(v_{y,E}+v_{y,C}\right)\left(v_{x,N}+v_{x,C}\right)-\frac{1}{2}\left(v_{y,S}+v_{y,SE}\right)\left(v_{x,S}+v_{x,C}\right)\right]$$

$$-\frac{\Delta t}{\Delta l}\left(\bar{p}_E-\bar{p}_C\right)+\frac{\nu\Delta t}{(\Delta l)^2}\left(v_{x,N}+v_{x,E}+v_{x,S}+v_{x,W}-4v_{x,C}\right)\qquad(8.97)$$

$$v_{y,C}^{n+1}=\frac{1}{4}\left[v_{y,N}+v_{y,E}+v_{y,S}+v_{y,W}\right]-\frac{\Delta t}{2\Delta l}\left[v_{y,N}^2-v_{y,S}^2\right]$$

$$-\frac{\Delta t}{2\Delta l}\left[\frac{1}{2}\left(v_{x,N}+v_{x,C}\right)\left(v_{y,E}+v_{y,C}\right)-\frac{1}{2}\left(v_{x,NW}+v_{x,W}\right)\left(v_{y,W}+v_{y,C}\right)\right]$$

$$-\frac{\Delta t}{\Delta l}\left(\bar{p}_N-\bar{p}_C\right)+\frac{\nu\Delta t}{(\Delta l)^2}\left(v_{y,N}+v_{y,E}+v_{y,S}+v_{y,W}-4v_{y,C}\right)\qquad(8.98)$$

Inserting the new velocity components in 8.96 we find

$$D_C^{n+1}=\frac{1}{4}\left(D_N^n+D_E^n+D_S^n+D_W^n\right)-\frac{\Delta t}{2(\Delta l)^2}S_C^n$$

$$-\frac{\Delta t}{(\Delta l)^2}\left(\bar{p}_N^n+\bar{p}_E^n+\bar{p}_S^n+\bar{p}_W^n-4\bar{p}_C^n\right)$$

$$+\frac{\nu\Delta t}{(\Delta l)^2}\left(D_N^n+D_E^n+D_S^n+D_W^n-4D_C^n\right)\qquad(8.99)$$

with

$$S_C\equiv\left(v_{x,E}^2-v_{x,C}^2-v_{x,W}^2+v_{x,WW}^2\right)+\left(v_{y,N}^2-v_{y,C}^2-v_{y,S}^2+v_{y,SS}^2\right)$$

$$+\frac{1}{2}\left(v_{y,E}+v_{y,C}\right)\left(v_{x,N}+v_{x,C}\right)-\frac{1}{2}\left(v_{y,S}+v_{y,SE}\right)\left(v_{x,S}+v_{x,C}\right)$$

$$-\frac{1}{2}\left(v_{x,NW}+v_{x,W}\right)\left(v_{y,C}+v_{y,W}\right)+\frac{1}{2}\left(v_{x,W}+v_{x,SW}\right)\left(v_{y,ES}+v_{y,SW}\right)\qquad(8.100)$$

Next we have to solve the Poisson equation 8.94. If the methods for doing this were without error, and if indeed all $D_{i,j}^n$ and $D_{i,j}^{n+1}$ were zero, we could simply write

$$\bar{p}_N + \bar{p}_E + \bar{p}_S + \bar{p}_W - 4\bar{p} = -S_C \qquad (8.101)$$

to compute the pressures at each time step. The Lax method by which we have produced the new velocities is conservative, meaning that (disregarding machine errors) it would fulfill the condition $D_{i,j}^n = 0$ at all times. However, the Poisson solver introduces an error in $p_{i,j}^{n+1}$ which makes $D_{i,j}^{n+1}$ depart from zero. To balance this we take into account these non-vanishing values of the divergence at time $t_{n+1}$ and write in place of 8.101

$$\bar{p}_N + \bar{p}_E + \bar{p}_S + \bar{p}_W - 4\bar{p}_C = -S_C + \frac{(\Delta l)^2}{4\Delta t}(D_N + D_E + D_S + D_W)$$
$$+ \nu(D_N + D_E + D_S + D_W - 4D_C) \qquad (8.102)$$

In this manner we can prevent a gradual accumulation of errors which would produce spurious compressibility effects in the flow.

For the pressure method to be stable, once again the conditions

$$\Delta t \leq \frac{\Delta l}{\sqrt{2}|v|_{max}} \quad \text{and} \quad \Delta t \leq \frac{1}{2}\frac{(\Delta l)^2}{\nu} \qquad (8.103)$$

must be met.

## 8.2.3    Free Surfaces: Marker-and-Cell Method (MAC)

Thus far we have assumed the liquid to reach up to the vessel walls at all sides. A barytropic liquid, however, is capable of spontaneously forming a free surface as a boundary against the "vacuum". In the MAC (*marker and cell*) method appropriate boundary conditions are introduced to handle such an open surface. The "marker" particles, which primarily serve to distinguish between liquid-filled and empty Euler cells, may also be utilized for the graphical representation of the shape of the liquid surface.

To integrate the hydrodynamic equations

$$\frac{\partial \mathbf{v}}{\partial t} = -\nabla\bar{p} - (\mathbf{v} \cdot \nabla)\mathbf{v} + \nu\nabla^2\mathbf{v} + \mathbf{g} \qquad (8.104)$$
$$\nabla \cdot \mathbf{v} = 0 \qquad (8.105)$$

one makes use of any of the foregoing techniques – the pressure method seems most popular in this context. However, each Euler cell now contains marker particles moving along according to the simple law $\mathbf{r}^{n+1} = \mathbf{r}^n + \mathbf{v}^n\Delta t$, where $\mathbf{v}^n$ is a particle velocity whose value is determined by interpolation, with suitable weights, from the velocities $v_x, v_y$ in the adjacent Euler cells [HARLOW 65].

The salient point here is the treatment of the Eulerian cells that constitute the free surface. There are four possible types of such interfacial cells. Figure 8.6 shows these four kinds of cells and the respective boundary conditions pertaining to the velocity components $v_x, v_y$. The boundary conditions for the pressure are the same in all cases: $p = p_{vac}$, where $p_{vac}$ is the "external" pressure in the empty Euler cells.

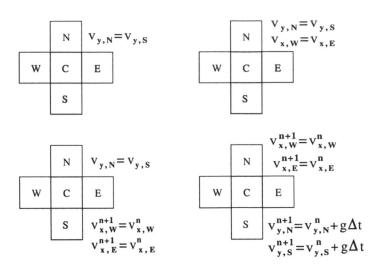

Figure 8.6: MAC method: the 4 types of surface cells and the appropriate boundary conditions for $v_x, v_y$ (see POTTER)

# 8.3 Lattice Gas Models for Hydrodynamics

## 8.3.1 Lattice Gas Cellular Automata

Cellular automata are one- or two-dimensional bit patterns that evolve in (discrete) time according to certain simple rules. The classic example is provided by John H. Conway's famous computer game "Life", in which each pixel on a screen is to be set or erased depending on the status of the neighboring pixels [EIGEN 82]. Informatics [WOLFRAM 86], evolution theory, and the mathematical theory of complexity [WOLFRAM 84] were quick to acquire this discretized representation of reality for their respective purposes. The following more physical application is just a kind of footnote to the broad theme of cellular automata.

Hardy, Pomeau and de Pazzis were the first to suggest a model representing a two-dimensional flow field in terms of bit patterns. Their "HPP model" works as follows [HARDY 73]:

A two-dimensional region is once more depicted by a grid of Eulerian cells, or "points". Each grid point $(i, j)$ may be populated by up to four "particles" whose velocities must point into different directions of the compass. The absolute value of the velocity is always $v = 1$.

Thus the number of possible "states" of a grid point is $2^4 = 16$. An economical way to describe the state of the grid point $(i, j)$ is to define a 4-bit (or half-byte) computer word $a_{i,j}$ representing the "empty" or "full" status of the compass directions E,N,W,S by one bit each (see Fig. 8.7). However, in many applications it is advantageous to combine the bits referring to the same direction at several successive grid points into one computer word. For example, in a 16 × 16 grid each

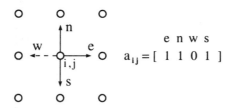

Figure 8.7: HPP model

| $e_0$ $n_0$ $w_0$ $s_0$ | $e_1$ $n_1$ $w_1$ $s_1$ |
|---|---|
| $e_2$ $n_2$ $w_2$ $s_2$ | $\cdots$ |
| | |
| | $\vdots$ |
| | $\cdots$ |
| $e_{30}$ $n_{30}$ $w_{30}$ $s_{30}$ | $e_{31}$ $n_{31}$ $w_{31}$ $s_{31}$ |

Figure 8.8: Storage methods in the HPP model

compass direction would be described by a set of 32 words of 1 byte each (see Fig. 8.8).

The state of the entire grid at time $t_{n+1}$ follows from the configuration at time $t_n$ according to a deterministic rule which is comprised of two substeps, *free flight* and *scattering*. In the free flight phase each particle moves on by one vertex in its direction of flight. In the example of Figure 8.8 each "north" bit in the second row (i.e. the bits in words $n_2$ and $n_3$), if it had value 1, would be reset to 0, while the respective bit above (in words $n_0$ and $n_1$) would be set to 1. Similar translations take place for the bit elements of the "south" words, while the 1-bits within the "east" and "west" words are right- and left-shifted, respectively, by one position.

In most programming languages logical operations may be performed not only with logical variables consisting of single bits, but also with byte-words or even integers made up of several bytes. In the above example the new word $n'_0$ could be computed as

$$n'_0 = n_0 \vee n_2 \tag{8.106}$$

with $\vee$ denoting the bitwise *or*-operation.

Analogous commands apply to the *s*-words. The compass directions *e* and *w* have to be handled, with this storing arrangement, in a bit-by-bit manner. However, nothing prevents us from combining the east and west bits column-wise; the translation may then be formulated as simply, and computed as speedily, as for north and south.

Figure 8.9: Scattering law for the HPP model

Obviously, one has to invent some plausible procedure for those bits that encounter any of the boundaries; there may be a law of reflection, or a periodic
boundary type rule. For example, if the grid is meant to describe the flow field
in the interior of a horizontal tube, it will make sense to decree that all $n$-bits in
the top row are to be transformed into $s$-bits before the translation takes place:
this is reflection. At the left and right borders one may assume periodic boundary
conditions. Reflection laws may also be used to outline the shapes of any obstacles
that may be present within the flow region.

Periodic boundary conditions will preserve momentum and energy exactly,
while in the presence of reflexion the conservation laws can hold only on the average.

Now for the second step, *scattering*. If after the translation step a grid point
is inhabited by two particles, its state is changed according to the rule depicted in
Fig. 8.9. In all other cases the state remains unaltered. Momentum and energy
are conserved by this scattering rule. We may write the HPP scattering rule in a
concise, computer-adapted way as follows:

$$a'_{i,j} = \{e \oplus u,\, n \oplus u,\, w \oplus u,\, s \oplus u\} \tag{8.107}$$

where $a_{i,j} \equiv \{e, n, w, s\}$ is the state of grid point $(i, j)$ before scattering (but after
translation), and

$$u \equiv [(e \oplus n) \odot (w \oplus s)] \odot [e \oplus (\neg w)] \tag{8.108}$$

By $\odot$, $\oplus$ and $\neg$ we denote the logical operators *and, exclusive or*, and *not*. ($\oplus$ differs
from $\vee$ in that $1 \oplus 1 = 0$.) Instead of using these operators (and the respective
computer commands) one may store the set of scattering rules in terms of a lookup
table.

Primitive as this model may seem when compared to the usual description
of the flow field, it proves to be quite relevant for hydrodynamics [FRISCH 86,
WOLFRAM 86B]. The momentary population number at a grid point defines a
density at that position, and the sum of velocities at $(i, j)$ may be interpreted as
a local velocity density in a fluid. By analyzing the foregoing "rules of the game"
in a spatially and temporally coarse-grained manner one obtains for the *averaged*
dynamics of mass and velocity very suggestive formulae that closely resemble the
continuity and Navier-Stokes equations. The important practical point is that
in simulating a system by the above rules only logical operations between logical
or integer variables need be performed. Such calculations are much faster than

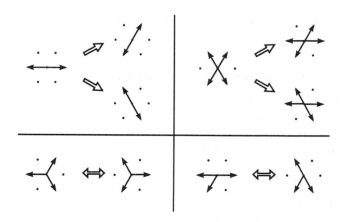

Figure 8.10: Scattering rules in the FHP model

the floating point operations needed for integrating the differential equations of hydrodynamics.

The still rather crude HPP model may be improved by the introduction of hexagonal cells in place of the simple quadratic lattice. In this "FHP model", thus named after the authors Frisch, Hasslacher, and Pomeau [FRISCH 86], there are six possible flight directions per grid point – and an accordingly larger number of scattering rules (see Figure 8.10). Further refinements of the model make allowance for the possibility of particles at rest, which makes for a still richer microdynamics.

The primary advantage of the FHP model over HPP is the fact that it guarantees rotational symmetry of the flow distribution, in spite of the discretization of particle velocities. In this respect, three-dimensional models require some more sophistication: in order to retain rotational symmetry, a *four*-dimensional face centered hypercubic (FCHC) lattice is set up and used in the propagation and collision steps. The results are then mapped onto three dimensions following a rule due to [D'HUMIÈRES].

It should be noted that in the basic HPP and FHP models the particles lose their identity in the process of scattering (see Figs. 8.9). It seems therefore that one cannot determine single particle properties, like velocity autocorrelations, by such simulations. However, it is always possible to "tag" some particles and augment the scattering law in such a way that in each scattering process the tags are passed on in a unique (be it random or deterministic) manner.

D. Frenkel has used such a procedure to study the long time behavior of the velocity autocorrelation function [FRENKEL 90, ERNST 91]. This is a *molecular* property all right, but at long times it will certainly be governed by hydrodynamic effects. It is a well underpinned tenet of kinetic theory that the "long time tail" of the velocity ACF should decay as $\propto t^{-d/2}$, where $d$ is the spatial dimension (2 or 3). However, the usual molecular dynamics simulations are not well suited to accurately study details of the long time behavior. By a two-dimensional FHP simulation with tagging, Frenkel et al. were able to produce unequivocal proof for the expected $t^{-1}$ decay.

## 8.3.2    The Lattice Boltzmann Method

Here is a promising development that arose from the lattice gas methods explained in the previous section. While the HPP and FHP calculations are very fast, they are also quite noisy. The fates of many single "particles" have to be followed, and many lattice sites have to be bundled together by a coarse-graining procedure, until reasonably clean flow patterns can be discerned. Both the computing speed and the noise are consequences of the discrete nature of the particle density in position-velocity space: at each grid point, any of the allowed discrete velocities $c_i$ may be taken by just one or no particle. In the Lattice Boltzmann method this latter rule is relaxed, and a floating point number is used to describe the degree to which each $(\mathbf{r}, \mathbf{c})$-cell is filled. It turns out that the loss in computing speed by the re-introduction of floating point arithmetic is more than compensated by the elimination of digital noise.

For easy notation the density, at time $t$, at position $\mathbf{r}$ and velocity $\mathbf{c}_i$ is denoted by $f_i(\mathbf{r}, t)$. Let the allowed velocity vectors point to each of the nearest neighbours on the lattice, and let their magnitudes be such that after one time step ($\Delta t = 1$, for simplicity) each particle will have arrived at the neighbouring site it aimed at. In a two-dimensional lattice square there are eight neighbours; including the possibility of particles at rest ($i = 0$), we are dealing with a set of nine numbers $[f_i(\mathbf{r}, t)\,,\ i = 0, 1, \ldots 8]$ at each grid point. The appropriate speeds are $|\mathbf{c}_0| = 0$, $|\mathbf{c}_i| = 1$ for the four directions along the grid axes, and $|\mathbf{c}_i| = \sqrt{2}$ for the diagonal directions. Similarly, in the 2-D hexagonal (FHP) lattice there are 6 nearest neighbours plus one rest particle. Three-dimensional models are again treated in the manner introduced by [D'HUMIÈRES] for lattice gas cellular automata.

The combined propagation-collision step may be written as

$$f_i(\mathbf{r} + \mathbf{c}_i, t + 1) = f_i(\mathbf{r}, t) + \Delta_i(f) \tag{8.109}$$

where $\Delta_i(f)$ denotes the increase or decrease of $f_i$ due to the collision process.

In early applications of the LB method the collision term was treated in the same way as in the underlying cellular automata model. This means that boolean operators were invoked to add or subtract inhabitants of a particular position-velocity cell. Around 1992 it was realized by two groups of authors ([QUIAN 92, CHEN 91]) that the LB model may be regarded as a representation of the Navier-Stokes equations, independent of the lattice gas cellular automata which precursed it. Following this train of thought, new approximations for the collision term were developed, and it turned out that the *single time relaxation* expression

$$\Delta_i(f) = -\frac{f_i - f_i^{eq}}{\tau} \tag{8.110}$$

is sufficient to reproduce Navier-Stokes dynamics. Here, $1/\tau$ is a suitable relaxation rate, and $f_i^{eq}$ denotes an appropriate equilibrium distribution. In the case of the 2-D hexagonal lattice this distribution is [QUIAN 95, CHEN 94]

$$f_{i=1...6}^{eq} = \frac{\rho}{12} + \frac{\rho}{3}\mathbf{e}_i \cdot \mathbf{v} + \frac{2\rho}{3}(\mathbf{e}_i \cdot \mathbf{v})^2 - \frac{\rho}{6} \cdot \mathbf{v}^2 \qquad (8.111)$$

$$f_{i=0}^{eq} = \frac{\rho}{2} - \rho\mathbf{v}^2 \qquad (8.112)$$

where $\mathbf{e}_i$ is a unit vector along $\mathbf{c}_i$, and $\rho$ and $\mathbf{v}$ are the hydrodynamic density and flow velocity, respectively:

$$\rho(\mathbf{r},t) = \sum_i f_i(\mathbf{r},t) \qquad \rho\mathbf{v}(\mathbf{r},t) = \sum_i \mathbf{c}_i f_i(\mathbf{r},t) \qquad (8.113)$$

Both compressible and incompressible flow may be treated using the Lattice Boltzmann method. Applications range from basic research on the dynamics of vortices to applied studies of turbulent channel flow, or oil recovery from sandstone. A recent survey of methodological refinements and specific applications is [QUIAN 95].

## 8.4    Direct Simulation Monte Carlo / Bird method

This not very enlightening name denotes an extremely successful semi-deterministic technique for solving flow problems in gases. It was introduced by G. Bird [BIRD 94] and further developed by, among others, K. Nanbu [NANBU 83] and F. Alexander, A. Garcia and B. Alder [ALEXANDER 95]. Its main area of application used to be dilute gas flow, both in earthbound engineering and in space science; recently, however, it was shown that the method may be extended so as to apply also to dense gases.[ALEXANDER 95]

There is a large and growing literature on the DSMC method and its applications. The basic method may be acquired by one of the books that contain sample codes, [GARCIA 99, BIRD 94]. A recent review is [GARCIA 97]. Search the web for the current developments.

The basic idea of the DSMC method as applied to a dilute gas of hard spheres may be sketched as follows:

1. Divide the sample to be studied into cells of volume $V_c$, each containing $N_c \approx 20 - 40$ particles whose position and velocity vectors are given. The side length of the cells should be smaller than but of the order of the mean free path. Boundary conditions appropriate to the problem at hand are defined, the most simple ones being specular (reflecting wall) and periodic boundaries. A time step $\Delta t$ smaller than the typical intercollision time is assumed.

2. Propagate all particles along their individual velocities according to $\mathbf{r}_i \rightarrow \mathbf{r}_i + \mathbf{v}_i \Delta t$, applying the respective boundary conditions.

3. Within each cell, draw $M_c$ pairs of particles $(i, j)$ that are to undergo a collision. As this step is at the heart of the method, we have to elaborate a bit.

   (a) First of all, the probability of a pair $(i, j)$ to collide is linked only to their relative speed $v_{ij} \equiv |\mathbf{v}_j - \mathbf{v}_i|$ and *not* to their positions. The argument for this is that all particles in one cell are within free path range of each other. The probability for the pair $(i, j)$ to collide is thus simply proportional to the relative speed: $p_c(i, j) \propto v_{ij}$. Recalling the rejection method of Section 3.2.4 we can easily see how to draw pairs $(i, j)$ in accordance with this probability density: assuming the maximum of $v_{ij}$ for all pairs in the cell to be known, draw a random number $\xi$ from a uniform distribution in $[0, 1]$ and compare it to $v_{ij}/v_{max}$. However, calculating $v_{max}$ would amount to the expensive scanning of all pairs of particles in the cell. The standard procedure therefore is to use an *estimated* value of $v'_{max}$. If that value is larger than the actual $v_{max}$, the density $p_{ij}$ is still sampled correctly but with a slightly lower efficiency.

   (b) The total number of collision pairs to be sampled in a cell during one time step is determined as follows. For a gas of hard spheres with diameter $d$ the average number of pair collisions within the cell is

   $$M_c \equiv Z V_c \Delta t = \frac{\rho^2 \pi d^2 \langle v_{rel} \rangle}{2} V_c \Delta t \qquad (8.114)$$

   where $Z$ is the kinetic collision rate per unit volume, $\rho = N_c/V_c$ is the number density, and $\langle v_{rel} \rangle$ is the average relative speed. In order to have $M_c$ trial pairs survive the rejection procedure of step 3a we have to sample

   $$M_{trial} \equiv M_c \frac{v'_{max}}{\langle v_{rel} \rangle} = \frac{\rho^2 \pi d^2 v'_{max}}{2} V_c \Delta t \qquad (8.115)$$

   collisions.

4. Having identified a pair of collision partners $(i, j)$, perform the actual collision. Since the post-collision velocities are determined by the impact parameter which is unknown, they must be sampled in a physically consistent way. In the hard sphere case this is most easily done by assuming an isotropic distribution of the *relative* velocity $\mathbf{v}_{ij}$ after the collision. Since the relative speed $|\mathbf{v}_{ij}|$ remains unchanged, the problem is reduced to sampling a uniformly distributed unit vector. Marsaglia's recipe may be used for this (see Figure 3.15).

5. Return to step 2.

# Appendixes

# Appendix A

# Machine Errors

This book is about algorithms, not machines. Nevertheless we will here display a few basic truths about the internal representation of numbers in computers. Keeping in mind such details often helps to keep the ubiquitous roundoff errors small.

In a generic 32-bit machine a *real* number is stored as follows:

| $\pm$ | e (exponent; 8 bits) | m (mantissa; 23 bits) |
|---|---|---|

or, in a more usual notation,
$$x = \pm m \cdot 2^{e - e_0}$$

- The mantissa $m$ is *normalized*, i.e. shifted to the left as far as possible, such that there is a 1 in the first position; each left-shift by one position makes the exponent $e$ smaller by 1. (Since the leftmost bit of $m$ is then known to be 1, it need not be stored at all, permitting one further left-shift and a corresponding gain in accuracy; $m$ then has an effective length of 24 bits.)

- The *bias* $e_0$ is a fixed, machine-specific *integer* number to be added to the "actual" exponent $e - e_0$, such that the stored exponent $e$ remains positive.

EXAMPLE: With a *bias* of $e_0 = 151$ (and keeping the high-end bit of the mantissa) the internal representation of the number 0.25 is, using $1/4 = (1 \cdot 2^{22}) \cdot 2^{-24}$ and $-24 + 151 = 127$,

$$\frac{1}{4} = \boxed{+ \mid 127 \mid 1\,0\,0 \ldots 0\,0}$$

Before any addition or subtraction the exponents of the two arguments must be equalized; to this end the *smaller* exponent is increased, and the respective mantissa is right-shifted (decreased). All bits of the mantissa that are thus being "expelled" at the right end are lost for the accuracy of the result. The resulting

| + | 35 | 1 1 1 ... 1 1 1 |

$-$  | + | 35 | 1 1 1 ... 1 1 0 |

$=$  | + | 35 | 0 0 0 ... 0 0 1 |

$=$  | + | 14 | 1 0 0 ... 0 0 0 |

Figure A.1: Subtraction of two almost equal numbers

error is called *roundoff error*. By *machine accuracy* we denote the smallest number that, when added to 1.0, produces a result $\neq 1.0$. In the above example the number $2^{-22} \equiv 2.38 \cdot 10^{-7}$, when added to 1.0, would just produce a result $\neq 1.0$, while the next smaller representable number $2^{-23} \equiv 1.19 \cdot 10^{-7}$ would leave not a rack behind:

1.0  | + | 129 | 1 0 0 ... 0 0 |

$+2^{-22}$  | + | 107 | 1 0 0 ... 0 0 |

$=$  | + | 129 | 1 0 0 ... 0 1 |

but

1.0  | + | 129 | 1 0 0 ... 0 0 |

$+2^{-23}$  | + | 106 | 1 0 0 ... 0 0 |

$=$  | + | 129 | 1 0 0 ... 0 0 |

A particularly dangerous situation arises when two almost equal numbers have to be subtracted. Such a case is depicted in Figure A.1. In the last (normalization) step the mantissa is arbitrarily filled up by zeros; the uncertainty of the result is 50%.

There is an everyday task in which such small differences may arise: solving the quadratic equation $ax^2 + bx + c = 0$. The usual formula

$$x_{1,2} = \frac{-b \pm \sqrt{b^2 - 4ac}}{2a} \tag{A.1}$$

will yield inaccurate results whenever $ac \ll b^2$. Since in writing a program one must always provide for the worst possible case, it is recommended to use the equivalent but less error-prone formula

$$x_1 = \frac{q}{a}, \qquad x_2 = \frac{c}{q} \tag{A.2}$$

with

$$q \equiv -\frac{1}{2}\left[b + sgn(b)\sqrt{b^2 - 4ac}\right] \qquad (A.3)$$

EXERCISE: Assess the machine accuracy of your computer by trying various negative powers of 2, each time adding and subtracting the number 1.0 and checking whether the result is zero.

# Appendix B

# Discrete Fourier Transformation

## B.1  Fundamentals

We are using the convention

$$\tilde{f}(\nu) = \int_{-\infty}^{\infty} f(t)\, e^{2\pi\, i\nu t}\, dt\,, \quad f(t) = \int_{-\infty}^{\infty} \tilde{f}(\nu)\, e^{-2\pi\, i\nu t}\, d\nu \tag{B.1}$$

Assume that the function $f(t)$ is given only at discrete, equidistant values of its argument:

$$f_k \equiv f(t_k) \equiv f(k\,\Delta t) \quad k = \ldots -2, -1, 0, 1, 2, \ldots \tag{B.2}$$

The reciprocal value of the time increment $\Delta t$ is called *sampling rate*. The higher the sampling rate, the more details of the given function $f(t)$ will be captured by the table of discrete values $f_k$. This intuitively evident fact is put in quantitative terms by *Nyquist's theorem*: if the Fourier spectrum of $f(t)$,

$$\tilde{f}(\nu) \equiv \int_{-\infty}^{\infty} f(t) e^{2\pi i\nu t}\, dt \tag{B.3}$$

is negligible for frequencies beyond the *critical* (or *Nyquist*) frequency

$$\pm\nu_0 \equiv \pm\frac{1}{2\Delta t} \tag{B.4}$$

then $f(t)$ is called a *band-limited process*. Such a process is completely determined by its sampled values $f_k$. The formula that permits the reconstruction of $f(t)$ from the sampled data reads

$$f(t) = \sum_{k=-\infty}^{\infty} f_k \frac{\sin[2\pi\nu_0(t - k\Delta t)]}{2\pi\nu_0(t - k\Delta t)} \tag{B.5}$$

(In contrast, if $f(t)$ is not band-limited, sampling with finite time resolution results in "mirroring in" the outlying parts of the spectrum from beyond $\pm\nu_0$, superposing them on the correct spectrum. In signal processing this effect is known as "aliasing".)

Let us assume now that a *finite* set of sampled values is given:

$$f_k, \quad k = 0, 1, \ldots N - 1 \tag{B.6}$$

and let $N$ be an even number. Define discrete frequencies by

$$\nu_n \equiv \frac{n}{N\Delta t}, \quad n = -\frac{N}{2}, \ldots, 0, \ldots, \frac{N}{2} \tag{B.7}$$

(The $\nu_n$ pertaining to $n = N/2$ is again the Nyquist frequency.) Then the Fourier transform of $f(t)$ at some frequency $\nu_n$ is given by

$$\tilde{f}(\nu_n) \approx \Delta t \sum_{k=0}^{N-1} f_k e^{2\pi i \nu_n t_k} = \Delta t \sum_{k=0}^{N-1} f_k e^{2\pi i k n/N} \tag{B.8}$$

Thus it makes sense to define the *discrete Fourier transform* as

$$\boxed{\begin{array}{c} F_n \equiv \displaystyle\sum_{k=0}^{N-1} f_k e^{2\pi i k n/N} \\[4pt] \text{with } N \text{ even, and } n = 0, \pm 1, \ldots, N/2 \end{array}} \tag{B.9}$$

According to B.8 the Fourier transform proper is just $\tilde{f}(\nu_n) \approx \Delta t\, F_n$.

From the definition of $F_n$ it follows that $F_{-n} = F_{N-n}$. We make use of this periodicity to renumber the $F_n$ such that $n$ runs from 0 to $N-1$ (instead of $-N/2$ to $N/2$):

$$-\tfrac{N}{2}, \ -\tfrac{N}{2}+1, \quad \ldots \quad 0, \quad \ldots \quad \tfrac{N}{2}-1, \quad \tfrac{N}{2}, \quad -\tfrac{N}{2}+1, \quad \ldots \quad -1$$

$$\Longrightarrow \qquad\qquad 0, \quad \ldots \quad \tfrac{N}{2}-1, \quad \pm\tfrac{N}{2}, \quad \tfrac{N}{2}+1, \quad \ldots \quad N-1$$

With this indexing convention the back transformation may be conveniently written

$$f_k = \frac{1}{N} \sum_{n=0}^{N-1} F_n e^{-2\pi i k n/N} \tag{B.10}$$

## B.2    Fast Fourier Transform (FFT)

If we were to use the definition B.9 "as is" to calculate the discrete Fourier transform, we would have to perform some $N^2$ operations. Cooley and Tukey (and before them Danielson and Lanczos; see [PRESS 86]) have demonstrated how, by smart handling of data, the number of operations may be pushed down to $\approx N \log_2 N$. Note that for $N = 1000$ this is an acceleration of $100 : 1$. Indeed, many algorithms of modern computational physics hinge on this possibility of rapidly transforming back and forth long tables of function values.

In the following it is always assumed that $N = 2^m$. If $N$ is not a power of 2, simply "pad" the table, putting $f_k = 0$ up to the next useful table length. Defining

$$W_N \equiv e^{2\pi i / N} \tag{B.11}$$

we realize that $W_N^2 = W_{N/2}$ etc. The discrete Fourier transform is therefore

$$F_N = \sum_{k=0}^{N-1} W_N^{nk} f_k \tag{B.12}$$

$$= \sum_{l=0}^{N/2-1} W_{N/2}^{nl} f_{2l} + W_N^n \sum_{l=0}^{N/2-1} W_{N/2}^{nl} f_{2l+1} \tag{B.13}$$

$$\equiv F_n^e + W_N^n F_n^o \tag{B.14}$$

where the indices $e$ and $o$ stand for "even" and "odd". Next we treat each of the two terms to the right of B.14 by the same pattern, finding

$$F_n^e = F_n^{ee} + W_{N/2}^n F_n^{eo} \tag{B.15}$$

$$F_n^o = F_n^{oe} + W_{N/2}^n F_n^{oo} \tag{B.16}$$

By iterating this procedure $m = \log_2 N$ times we finally arrive at terms $F_n^{(\ldots)}$ that are identical to the given table values $f_k$.

EXAMPLE: Putting $N = 4$ we have $W_4 \equiv exp[2\pi i/4]$ and

$$F_n = \sum_{k=0}^{3} W_4^{nk} f_k \quad n = 0, \ldots 3 \tag{B.17}$$

$$= \sum_{l=0}^{1} W_2^{nl} f_{2l} + W_4^n \sum_{l=0}^{1} W_2^{nl} f_{2l+1} \tag{B.18}$$

$$\equiv F_n^e + W_4^n F_n^o \tag{B.19}$$

$$= F_n^{ee} + W_2^n F_n^{eo} + W_4^n \left[ F_n^{oe} + W_2^n F_n^{oo} \right] \tag{B.20}$$

$$= f_0 + W_2^n f_2 + W_4^n \left[ f_1 + W_2^n f_3 \right] \tag{B.21}$$

Thus the correspondence between the table values $f_k$ and the terms $F_n^{ee}$ etc. is as follows:

|          | *ee* | *eo* | *oe* | *oo* |
|----------|------|------|------|------|
| $\longrightarrow$ | 0    | 2    | 1    | 3    |

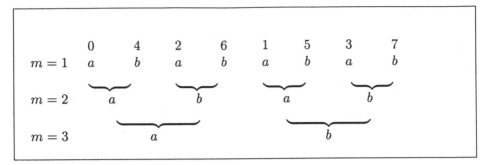

Figure B.1: Decimation for $N = 8$

EXERCISE: Demonstrate that a similar analysis as above leads for $N = 8$ to the correspondences

| $eee$ | $eeo$ | $eoe$ | $eoo$ | $oee$ | $oeo$ | $ooe$ | $ooo$ |
|---|---|---|---|---|---|---|---|
| 0 | 4 | 2 | 6 | 1 | 5 | 3 | 7 |

$\longrightarrow$

It is easy to see that this correspondence is reproduced by the following rule: 1) put $e \leftrightarrow 0$ and $o \leftrightarrow 1$, such that $eeo \leftrightarrow 001$ etc.; 2) reverse the bit pattern thus obtained and interpret the result as an integer number: $eeo \leftrightarrow 001 \leftrightarrow 100 = 4$. In other words, arrange the table values $f_k$ in bit-reversed order. (For example, $k = 4$ is at position 1 since $4 = 100 \longrightarrow 001 = 1$.)

The correctly arranged $f_k$ are now combined in pairs according to B.21. The rule to follow in performing this "decimation" step is sketched in Fig. B.1. On each level ($m$) the terms $a, b$ are combined according to

$$a + W_{2m}^n b \qquad\qquad (B.22)$$

It is evident that the number of operations is of order $N \log_2 N$.

Further details of the method, plus sample programs in Pascal, Fortran, or C are given in [PRESS 86].

EXERCISE: Sketch the pattern of Figure B.1 for $N = 4$ and perform the "decimation". Compare your result to equ. B.21.

# Bibliography

[ABRAMOWITZ 65] Abramowitz, M., Stegun, I. A., eds.: Handbook of Mathematical Functions. Dover, New York, 1965.

[ALDER 57] Alder, B. J. and Wainwright, T. E., J. Chem. Phys. 27 (1957) 1208.

[ALDER 67] Alder, B. J., and Wainwright, T. E., Phys. Rev. Lett. 18 (1967) 988.

[ALEXANDER 95] Alexander, F. J., Garcia, A. L., and Alder, B. J., Phys. Rev. Lett. 74 (1995) 5212.

[ALLEN 90] Allen, M. P., and Tildesley, D. J.: Computer Simulation of Liquids. Oxford University Press, 1990.

[BARNETT 86] Barnett, R. N., Reynolds, P. J., and Lester, W. A., J. Chem. Phys. 84 (1986) 4992.

[BINDER 87] Binder, K.: Applications of the Monte Carlo Method in Statistical Physics. Springer, Berlin 1987.

[BINDER 92] Binder, K.: The Monte Carlo Method in Condensed Matter Physics. Springer, Berlin 1992.

[BIRD 94] Bird, G. A.: Molecular Gas Dynamics and the Direct Simulation of Gas Flows. Oxford University Press.

[BRIGGS 94] Briggs, W. L.: A Multigrid Tutorial. Society for Industrial and Applied Mathematics (SIAM), Philadelphia, Pennsylvania 1987,1994.

[BROOKS 83] Brooks, B. R., Bruccoleri, R. E., Olafson, B. D., States, D. J., Swaminathan, S., and Karplus, M., CHARMM: A Program for Macromolecular Energy, Minimization, and Dynamics Calculations, J. Comp. Chem. 4 (1983) 187-217.

[CANDY 91] Candy, J., and Rozmus, W., J. Computat. Phys. 92 (1991) 230.

[CAR 85] Car, R., and Parrinello, M., Phys. Rev. Lett. 55 (1985) 2471.

[CAVENDISH WWW] Web site on Quantum Monte Carlo Simulations, Cavendish Laboratory, University of Cambridge: www.tcm.phy.cam.ac.uk/~mdt26/cqmc.html

[CEPERLEY 80] Ceperley, D. M., and Alder, B. J., Phys. Rev. Lett. 45/7 (1980) 566; Science 231 (1986) 555.

[CEPERLEY 88] Ceperley, D. M., and Bernu, B., J. Chem. Phys. 89 (1988) 6316.

[CEPERLEY 95] Ceperley, D. M., Path Integrals in the Theory of Condensed Helium. Rev. Mod. Phys. 67 (1995) 279.

[CEPERLEY 96] Ceperley, D. M., and Mitas, L., Quantum Monte Carlo Methods in Chemistry, in: Prigogine, I., and Rice, S. A. (Eds.): New Methods in Computational Quantum Mechanics. Advances in Chemical Physics, XCIII. John Wiley, New York 1996.

[CEPERLEY WWW] Web site of the National Center for Supercomputing Applications, dept. of Condensed Matter Physics: www.ncsa.uiuc.edu/Apps/CMP/

[CHANDLER 81] Chandler, D., and Wolynes, P. G., J. Chem. Phys. 74 (1981) 4078.

[CHEN 91] Chen, S., Chen, H., Martinez, D., and Matthaeus, W. H., Phys. Rev. Lett. 67 (1991) 3776.

[CHEN 94] Chen, S., Doolen, G. D., and Eggert, K. G., Los Alamos Science 22 (1994) 98.

[CICERO -44] Cicero, M. T.: De Fato, Book XX. Rome, 44 A.C. Quoted from: Cicero: On Fate & Boethius: The Consolation of Philosophy. Edited with Introduction, Translation and Commentary by R. W. Sharples. Aris & Phillips Ltd., Warminster, England.

[COKER 87] Coker, D. F., Berne, B. J., and Thirumalai, D., J. Chem. Phys. 86 (1987) 5689.

[COLDWELL 74] Coldwell, R. L., J. Computat. Phys. 14 (1974) 223.

[COOPER 89] Cooper, Necia G.: The beginning of the MC method. In: Cooper, N. G., ed.: From Cardinals to Chaos. Reflections on the Life and Legacy of Stanislaw Ulam. Cambridge University Press, New York 1989.

[D'HUMIÈRES] D'Humières, D., Lallemand, P., and Frisch, U., Europhys. Lett. 2 (1986) 1505.

[DORFMAN 72] Dorfman, J. R., and Cohen, E. G. D., Phys. Rev. A6 (1972) 776.

[EIGEN 82] Eigen, Manfred, and Winkler, Ruthild: Das Spiel - Naturgesetze steuern den Zufall. Piper, Munich/Zurich 1982. (See also: Gardner, M., Scientific American, Oct. 1970 and Feb. 1971.)

[ENGELN 91] Engeln-Muellges, G., and Reutter, F.: Formelsammlung zur Numerischen Mathematik mit Standard-FORTRAN 77-Programmen. BI-Verlag, Mannheim, 1991.

[ERNST 91] Ernst, M. H., in: Hansen, J.-P., Levesque, D., and Zinn-Justin, J. (eds.): Liquides, Cristallisation et Transition Vitreuse. Les Houches, Session LI. North-Holland, Amsterdam 1991.

[EVANS 86] Evans, D. J.: Nonequilibrium molecular dynamics. In: Ciccotti, G., and Hoover, W. G., eds.: Molecular dynamics simulation of statistical-mechanical systems. (Proceedincs, Intern. School of Physics "Enrico Fermi"; course 97) North-Holland, Amsterdam 1986.

[EWALD 21] Ewald, P. P., Ann. Phys. 64 (1921) 253.

[FRENKEL 90] Frenkel, D., in: Van Beijeren, H.: Fundamental Problems in Statistical Mechanics VII. North-Holland, Amsterdam 1990.

[FRISCH 86] Frisch, U., Hasslacher, B., and Pomeau, Y., Phys. Rev. Lett. 56 (1986) 1505.

[GALLI 90A] Galli, G., and Parrinello, M., J. Phys.: Condensed Matter 2 (1990) SA227.

[GALLI 90B] Galli, G. et al., Phys. Rev. B 42/12 (1990) 7470.

[GARCIA 97] Garcia, A. L., and Baras, F.: Direct Simulation Monte Carlo: Novel Applications and New Extensions. In: Proceedings of the Third Workshop on Modelling of Chemical Reaction Systems, Heidelberg (1997). CD-ROM ISBN 3-932217-00-4, or download from "www.wenet.net/~algarcia/Pubs/".

[GARCIA 99] Garcia, A. L.: Numerical Methods for Physics. Prentice Hall, New Jersey, 1999.

[GEAR 66] Gear, C. W., Argonne Natl. Lab. Report ANL-7126 (1966).

[GEAR 71] Gear, C. W.: Numerical Initial Value Problems in Ordinary Differential Equations. Prentice-Hall, New Jersey 1971.

[GINGOLD 77] Gingold, R. A., and Monaghan, J. J., Mon. Not. Roy. Astron. Soc. 181 (1977) 375.

[GOLDBERG 89] Goldberg, D. E.: Genetic Algorithms in Search, Optimization and Machine Learning. Addison-Wesley, Reading, MA 1989.

[GOLDSTEIN 80] Goldstein, H.: Classical Mechanics; Second Edition. Addison-Wesley, Reading, Mass., 1980.

[GUTBROD 99] Gutbrod, F.: New trends in pseudo-random number generation, in: Stauffer D. (Ed.), Ann. Rev. Comp. Phys. VI, World Scientific, 1999.

[HANSEN 69] Hansen, J.-P., and Verlet, L., Phys. Rev. 184 (1969) 151.

[HANSEN 86] Hansen, J.-P., and McDonald, I. R.: Theory of Simple Liquids; 2nd edition. Academic Press, London 1986.

[HARDY 73] Hardy, J., Pomeau, Y., and de Pazzis, O., J. Math Phys. 14 (1973) 1746; Phys. Rev., A 13 (1976) 1949.

[HARLOW 65] Harlow, F. H., and Welch, J. E., Phys. Fluids 8 (1965) 2182.

[HEHRE 86] Hehre, W. J., Radom, L., Schleyer, P. v. R., and Pople, J. A.: Ab initio molecular orbital theory. Wiley, New York 1986.

[HELLER 75] Heller, E. J., J. Chem. Phys. 62 (1975) 1544.

[HELLER 76] Heller, E. J., J. Chem. Phys. 64 (1976) 63.

[HERRERO 95] Herrero, C. P., Ramirez, R., Phys. Rev. B 51 (1995) 16761.

[HEYES 86] Heyes, D. M., Mol. Phys. 57/6 (1986) 1265.

[HOCKNEY 70] Hockney, R. W.: The potential calculation and some applications, in: Alder, B., Fernbach, S., Rotenberg, M. (eds.): Methods in Computational Physics, Vol. 9, Plasma Physics. Academic Press, New York/London, 1970.

[HOCKNEY 81] Hockney, R. W., and Eastwood, J. W.: Computer Simulation Using Particles. McGraw-Hill, New York 1981.

[HOLIAN 87] Holian, B. L., Hoover, W. G., Posch, H. A., Phys. Rev. Lett. 59/1 (1987) 10.

[HOLIAN 95] Holian, B. L., Voter, A. F., and Ravelo, R., Phys. Rev. E 52 (1995) 2338.

[HONERKAMP 91] Honerkamp, J.: Stochastische Dynamische Systeme. Verlag Chemie, Weinheim 1991.

[HOOVER 68] Hoover, W. G., and Ree, F. H., J. Chem. Phys. 49 (1968) 3609.

[HOOVER 91] Hoover, W. G.: Computational Statistical Mechanics. Elsevier, Amsterdam, Oxford, New York, Tokyo 1991.

[HOOVER 99] Hoover, W. G.: Time Reversibility, Computer Simulation, and Chaos. World Scientific, Singapore, New Jersey, London, Hong Kong 1999.

[HUBER 88] Huber, D., and Heller, E. J., J. Chem. Phys. 89/8 (1988) 4752.

[IVANOV 00] Ivanov, D., Doctoral Dissertation, University of Vienna, 2000.

[JAMES 90] James, F.: A Review of Pseudorandom Generators. CERN-Data Handling Division, Rep. No. DD/88/22, 1988.

[JENNEWEIN 00] Jennewein, Th., Achleitner, U., Weihs, G., Weinfurter, H., and Zeilinger, A., Rev. Sci. Instr. 71/4 (2000) 1675.

[JONES 1711] Jones, William, ed.: Analysis per Quantitatum Series, Fluxiones, ac Differentias. London 1711.

[KALOS 74] Kalos, M. H., Levesque, D., and Verlet, L., Phys. Rev. A 138 (1974) 257.

[KALOS 86] Kalos, M. H., and Whitlock, P. A.: Monte Carlo Methods. Wiley, New York 1986.

[KASTENMEIER 86] Kastenmeier, Th., and Vesely, F. J., Robotica 14 (1996) 329.

[KIRKPATRICK 81] Kirkpatrick, S., and Stoll, E. P., J. Comp. Phys. 40 (1981) 517.

[KIRKPATRICK 83] Kirkpatrick S., Gelatt, C. D., Jr., and Vecchi, M. P., Science 220 (1983) 671.

[KNAUP 99] Knaup, M., Reinhard, P.-G., and Toepfer, Ch., Contrib. Plasma Phys. 39 (1999) 57.

[KNUTH 69] Knuth, D. E.: The Art of Computer Programming. Addison-Wesley, Reading, Massachusetts, 1969.

[KOHLER 72] Kohler, F., Findenegg, G. H., Fischer, J., Posch, H., and Weissenboeck, F.: The Liquid State. Verlag Chemie, Weinheim 1972.

[KOHN 65] Kohn, W., and Sham, L. J., Phys. Rev. 140 (1965) A1133.

[KOLAR 89] Kolar, M., and Ali, M. K., J. Chem. Phys. 90/2 (1989) 1036.

[KOONIN 85] Koonin, S. E.: Computational Physics. Benjamin, New York 1985.

[KUBO 71] Ryogo Kubo: Statistical Mechanics. North-Holland, Amsterdam/London 1971.

[LANDAU 62] Landau, L. D., and Lifschitz, E. M.: Mechanik. Berlin 1962.

[LANKFORD 90] Lankford, J., and Slavings, R. L., Physics Today, March 1990, p.58.

[LEVESQUE 69] Levesque, D., and Verlet, L., Phys. Rev. 182 (1969) 307; Phys. Rev. A2/6 (1970) 2514; – and Kuerkijaervi, J., Phys. Rev. A7/5 (1973) 1690.

[LUCY 77] Lucy, L. B., The Astronomical Journal 82/12 (1977) 1013.

[MACKERRELL 98] MacKerell, A. D., Brooks, B., Brooks, C. L., Nilsson, L., Roux, B., Won, Y., and Karplus, M.: CHARMM: The Energy Function and Its Parameterization with an Overview of the Program, in: Schleyer, P. v. R. et al., editors: The Encyclopedia of Computational Chemistry, John Wiley & Sons, Chichester 1998.

[MAKRI 99] Makri, N., Ann. Rev. Phys. Chem. 50 (1999) 167.

[MARSAGLIA 72] Marsaglia, G., Ann. Math. Stat. 43/2 (1972) 645.

[MARSAGLIA 90] Marsaglia, G., and Zaman, A., Stat. & Probab. Letters 8 (1990) 35.

[MARTYNA 92] Martyna, G. J., Klein, M., and Tuckerman, M. J., Chem. Phys. 97 (1992) 2635.

[MARTINEZ 97] Martinez, T. J., Ben-Nun, M., Levine, R. D., J. Phys. Chem. 101 (1997) 6389.

[MAZUR 70] Mazur, P., and Oppenheim, I., Physica 50 (1970) 241.

[MAZZONE 99] Mazzone, A. M., in: Stauffer, D. (Ed.): Ann. Rev. Comp. Ph. VI, World Scientific, Singapore 1999.

[MCDONALD 74]  McDonald I. R., J. Phys. C: Sol. St. Ph., 7 (1974) 1225.

[MCKEOWN 87]  McKeown, P. K., and Newman, D. J.: Computational Techniques in Physics. Adam Hilger, Bristol 1987.

[METROPOLIS 49]  Metropolis, N., and Ulam, S., J. Amer. Statist. Assoc. 44 (1949) 335.

[METROPOLIS 53]  Metropolis, N. A., Rosenbluth, A. W., Rosenbluth, M. N., Teller, A. H., and Teller, E., J. Chem. Phys. 21 (1953) 1087.

[MEYERS 56]  Meyers, H. A. (ed.): Symposium on Monte Carlo Methods. Wiley, New York 1956.

[MONAGHAN 92]  Monaghan, J. J., Ann. Rev. Astron. Astrophysics 30 (1992) 543.

[MONAGHAN 89]  Monaghan, J. J., J. Comput. Phys. 82/1 (1989) 1.

[MORI 65]  Mori, H., Prog. theor. Phys. 34 (1965) 399.

[MULLER 58]  Muller, M. E., Math. Tables Aids Comp. 63 (1958) 167.

[NANBU 83]  Nanbu, K., J. Phys. Soc. Jpn. 52 (1983) 3382.

[NERI 88]  Neri, F., preprint, Dept. of Physics, Univ. of Maryland, 1988.

[NEUMANN 86]  Neumann, M., J. Chem. Phys. 85 (1986) 1567.

[NEWTON 1674]  Newton, Sir Isaac, quoted from Whiteside, D. T., ed.: The Mathematical Papers of Isaac Newton, Volume IV, 1674-1684. Cambridge University Press, 1971, p.6.

[NIEDERREITER 82]  Niederreiter, H., in: Grossmann, W., et al., eds.: Probability and Statistical Inference. Reidel, Dordrecht, 1982.

[NILSSON 90]  Nilsson, L. G., and Padrò, J. A., Mol. Phys. 71 (1990) 355.

[NOSÉ 91]  Nosé, Sh.: The development of Molecular Dynamics simulations in the 1980s. In: Yonezawa, F., Ed.: Molecular Dynamics Simulation. Springer, Berlin 1992.

[NUGENT 00]  Nugent, S., and Posch, H. A., Phys. Rev. E 62/4 (2000), to appear.

[OHNO 99]  Ohno, K., Esfarjani, K., and Kawazoe, Y.: Computational Materials Science – From Ab Initio to Monte Carlo Methods. Springer, Berlin 1999.

[PAPOULIS 81]  Papoulis, Athanasios: Probability, Random Variables and Stochastic Processes. McGraw-Hill International Book Company, 1981.

[PARRINELLO 84]  Parrinello, Rahman, J. Chem. Phys. 80 (1984) 860.

[POSCH 89]  Posch, H. A., and Hoover, W. G., Phys. Rev. A 39/4 (1989) 2175.

[POSCH 90]  Posch, H. A., Hoover, W. G., and Holian, B. L., Ber. Bunseng. Phys. Chem. 94 (1990) 250.

[POSCH 92] Posch, H. A., and Hoover, W. G., in: J. J. C. Texeira-Dias (ed.): Molecular Liquids – New Perspectives in Physics and Chemistry. Kluwer Academic, Netherlands 1992; p. 527.

[POSCH 97] Posch, H. A., and Hoover, W. G., Phys. Rev. E 55 (1997) 6803.

[POTTER 80] Potter, D.: Computational Physics. Wiley, New York 1980.

[PRESS 86] Press, W. H., Flannery, B. P., Teukolsky, S. A., and Vetterling, W. T.: Numerical Recipes – The Art of Scientific Computing. Cambridge University Press, New York 1986.

[QUIAN 92] Quian, Y. H., D'Humières, D., and Lallemand, P., Europhys. Lett. 17 (1992) 479.

[QUIAN 95] Quian, Y. H., Succi, S., and Orszag, S., Ann. Rev. Comput. Phys. III (1995) 195.

[RAHMAN 64] Rahman, A., Phys. Rev. 136/2A (1964) A405.

[RAHMAN 71] Rahman, A., Stillinger, F. H., J. Chem. Phys. 55/7 (1971) 3336.

[RAPAPORT 88] Rapaport, D. C.: Molecular Dynamics: A New Approach to Hydrodynamics? In: Landau, D. P., Mon, K. K., and Schuettler, H.-B., eds.: Springer Proceedings in Physics, Vol. 33: Computer Simulation Studies in Condensed Matter Physics. Springer, Berlin 1988.

[RUTH 83] Ruth, R. D., IEEE Trans. Nucl. Sci. NS-30 (1983) 2669.

[RYCKAERT 77] Ryckaert, J. P., Ciccotti, G., and Berendsen, H. J. C., J. Comp. Phys. 23 (1977) 327.

[SINGER 86] Singer, K., and Smith, W., Mol. Phys. 57/4 (1986) 761.

[SKINNER 85] Skinner, D. W., Moskowitz, J. W., Lee, M. A., Whitlock, P. A., and Schmidt, K. E., J. Chem. Phys. 83 (1985) 4668.

[SMITH 90] Smith, D. E., and Harris, C. B., J. Chem. Phys. 92 (1990) 1304.

[SMITH 96] Smith, W., and Forrester, T. R., J. Molec. Graphics 14 (1996) 136.

[SMITH WWW] Smith, W., and Forrester, T. R., downloadable software package DL_POLY: www.dl.ac.uk/TCS/Software/DL_POLY/

[STAUFFER 89] Stauffer, D., Hehl, F. W., Winkelmann, V., and Zabolitzky, J. G.: Computer Simulation and Computer Algebra. Springer, Berlin 1989.

[STOER 89] Stoer, J.: Numerische Mathematik. Springer, Berlin 1989.

[STOERMER 07] Størmer, C., Arch. Sci. Phys. Nat. Genève, 1907.

[STOERMER 21] Størmer, C., in: Congres International Mathematique Strasbourg. Toulouse 1921.

[SWOPE 82] Swope, W. C., Andersen, H. C., Berens, P. H., and Wilson, K. R., J. Chem. Phys. 76 (1982) 637.

[TAUSWORTHE 65] Tausworthe, R. C., Math. Comp. 19 (1965) 201.

[TOMASSINI 95] Tomassini, M.: A Survey of Genetic Algorithms. In: Ann. Revs. Comp. Phys., Vol. III, p. 87. World Scientific, Singapore 1995.

[ULAM 47] Ulam, S., Richtmeyer, R. D., Von Neumann, J., Los Alamos Nat. Lab. Sci. Rep. LAMS-551, 1947.

[VAN GUNSTEREN 84] Van Gunsteren, W. F., and Berendsen, H. J. C., GROMOS Software Manual. University of Groningen 1984.

[VERLET 67] Verlet, L., Phys. Rev. 159/1 (1967) 98; ibidem, 165/1 (1968) 201.

[VESELY 78] Vesely, F. J.: Computerexperimente an Fluessigkeitsmodellen. Physik Verlag, Weinheim 1978.

[VESELY 82] Vesely, F. J., J. Computat. Phys. 47/2 (1982) 291.

[VESELY 84] Vesely, F. J., Mol. Phys. 53 (1984) 505.

[VINEYARD 62] Vineyard, G. H., Proc. Intern. School of Physics Enrico Fermi, Course XVIII (1960). Academic Press, New York 1962.

[VITEK 89] Vitek, V., Srolovitz, D. J., eds.: Atomistic simulations of materials: Beyond pair potentials. Plenum, New York 1989.

[WENTZCOVICH 91] Wentzcovich, R. M., and Martins, J. L., University of Minnesota Supercomputer Institute Res. Rep. UMSI 91/13, 1991.

[WESSELING 92] Wesseling, P.: An Introduction to Multigrid Methods. John Wiley, New York 1992.

[WHITLOCK 79] Whitlock et al., Phys. Rev. B 19 (1979) 5598.

[WILKINSON 67] Wilkinson, J. H.: in Klerer, M. and Korn, G. A.(Eds.): Digital Computer User's Handbook, McGraw-Hill, New York 1967.

[WOLFRAM 84] Wolfram, S., Nature 311 (1984) 419.

[WOLFRAM 86] Wolfram, S.: Theory and Applications of Cellular Automata. World Scientific, 1986.

[WOLFRAM 86B] Wolfram, S.: J. Statist. Phys. 45/3-4 (1986) 471.

[YOSHIDA 93] Yoshida, H., Celest. Mech. Dynam. Astron. 56 (1993) 27.

[ZOPPI 91] Zoppi, M., and Neumann, M., Phys. Rev. B43 (1991) 10242.

# Index